The Price of Precaution and the Ethics of Risk

The International Library of Ethics, Law and Technology

VOLUME 6

For further volumes:
http://www.springer.com/series/7761

Christian Munthe

The Price of Precaution and the Ethics of Risk

Springer

Christian Munthe
Department of Philosophy, Linguistics
and Theory of Science
University of Gothenburg
Box 200, 40530 Gothenburg
Sweden
Christian.Munthe@gu.se

ISSN 1875-0044 e-ISSN 1875-0036
ISBN 978-94-007-1329-1 e-ISBN 978-94-007-1330-7
DOI 10.1007/978-94-007-1330-7
Springer Dordrecht Heidelberg London New York

Library of Congress Control Number: 2011926415

Printed on acid-free paper

Springer is part of Springer Science+Business Media (www.springer.com)

*I dedicate this book to my daughter, Saga –
born in the same year as I started this work
and celebrating her 11th birthday in the year
of its completion.*

Acknowledgements

The main bulk of the work on this book was undertaken within the research project *Nature and Precaution: Conceptual and Ethical Issues in Environmental Philosophy and Policy Making*, generously supported by the Swedish Research Council.

Jonas Gren took the pains of carefully reading and commenting on an extensive first draft, besides being an inspiring discussion partner during the process of writing. Sven-Ove Hansson once upon a time introduced me to the intersection of areas such as risk analysis, decision theory, normative ethics and political philosophy that forms the theoretical landscape that I have been working within. He has also been very generous of having me visiting his unit at the Royal Institute of Technology in Stockholm, as well as a number of other arenas, for various presentations and discussions. Several other people have assisted tremendously over the years by their willingness to discuss many of the issues addressed in this book and, in some cases, actual drafts of various sections: Petra Andersson, Richard Ashcroft, Bengt Brülde, Linus Broström, John Cantwell, Angus Dawson, Dan Egonsson, Ragnar Francén, Sören Halldén, Matti Häyry, Niklas Juth, Karsten Klint Jensen, Jan Lif, Pia Nykänen, Anders J Persson, Ingmar Persson, Johannes Persson, Martin Peterson, Wlodek Rabinowicz, Toni Rønnow-Rasmussen, Per Sandin, Lars Sandman and Anders Tolland. During the processing of the manuscript by Springer, three anonymous reviewers added further material for improvements. Needless to say, none of these people carry any responsibility for remaining flaws.

Contents

Chapter 1
Introduction

1.1 Background

The so-called *Precautionary Principle* (PP) has come to be a standard notion in environmental policy and debate. It appears frequently in important policy documents and international agreements, as well as in the environmental legislation of various countries. In pair with its sister notion of 'sustainable development', it has become *the* default piece of rhetoric and argumentation embraced by almost all sides in public discussions about environmental policy issues and controversies regarding the use of technology.

Whilst the exact historical timing of the emergence of the notion of PP may be debated,[1] the background of its *current popularity* seems comparably clear. The environmental problems of industrialised societies, highlighted in public discourse since the 1970s, have given rise to a widely shared sense of failure in the otherwise broadly embraced enlightment ideal of a rationally planned and scientifically based management of societal endeavours. Powered by (sometimes alleged) facts about pollution due to our means of production and transportation, exploitation of natural resources in our hunt for energy and materials, and an ever-growing mountain of waste left to future generations to care for, this failure seems particularly salient when it comes to our use of technology and its effect on the natural environment. We have all come to realise that, despite the best intentions and most considerate concern for human welfare, our efforts of improving the world may in the end go terribly wrong. The dangers we are trying to master may be exchanged for even more fearful scenarios created by these our own very efforts. The currently much highlighted threats due to fossil fuel emissions and climate change is the latest, and perhaps also the most serious, challenge of this sort. In effect, a principle of precaution urging us to reflect on the possibility of disaster beforehand and to proceed with caution in the light of such awareness may be seen as highly called for.

[1] For a brief overview of different views on this matter, see Sandin (2004, pp. 3–5).

C. Munthe, *The Price of Precaution and the Ethics of Risk*, The International Library of Ethics, Law and Technology 6, DOI 10.1007/978-94-007-1330-7_1,

Just as the historical roots of PP may be debated, its exact *route* of entrance onto the scene of contemporary environmental politics may also be a topic for discussion.[2] However, it seems unquestionable that the notion of PP would not carry the weight it does today had it not been for the reference to "the precautionary approach" made in principle 15 of the United Nation's *Rio Declaration* from 1992:

> In order to protect the environment, the precautionary approach shall be widely applied by States according to their capabilities. Where there are threats of serious or irreversible damage, lack of full scientific certainty shall not be used as a reason for postponing cost-effective measures to prevent environmental degradation.[3]

This formulation has most certainly played a key role in the inclusion and mention of PP in a host of other central environmental policy documents. Today, PP is spoken about or referred to both in various multilateral agreements and international law, as well as in policy statements and legislations of many individual countries.[4]

1.1.1 Diversity and Unclarity

So far, it has been possible to speak loosely about the general *notion* of PP. But if something more is to be said about and (much more importantly) *done with* this notion we must also be able to speak more specifically about *the* PP. However, this immediately proves to be quite a problematic venture.[5]

First of all, no one needs to look very hard at the just quoted formulation in order to realise that it is highly unclear in several respects. Among the queries that may be directed at the *Rio Declaration*'s version of PP are: What is to count as "a threat"? What is to count as "serious or irreversible damage"? What is meant by "full scientific certainty"? What determines whether or not there is a "lack" of such certainty? What is meant by the proviso that such lack "shall not be used as a reason"? What determines the extent to which some measure is "cost-effective"? What is meant by "environmental degradation" What is meant by the proviso that what is said only concerns "measures to prevent" such degradation? And so on.[6] Moreover, each of these queries may be answered in a multitude of mutually incompatible ways. In effect, what the *Rio Declaration* expresses, and the international community thereby has agreed on, is a long list of possible PPs that say quite different things, but that nevertheless are all in compliance with the requirements of the *Rio Declaration*.

Secondly, due to the overwhelming influence of the *Rio Declaration*, there is nowadays no end to the number of formulations and statements available that by someone or other has been dignified with the golden epithet of PP (and each of these carry unclarities similar to those of the PP of the *Rio Declaration*). This goes not

[2] Sandin (2004, pp. 4–5).

[3] *The Earth Summit*.

[4] See O'Riordan et al. (2001a, chapters 5–10); and Trouwborst (2002).

[5] Cf. Graham (2000).

[6] All of these unclarities, as well as several others, will be addressed throughout this book.

only for policy statements, political agreements and legislation, but also for a large number of claims made in debates on environmental policy, technology use and a host of other controversial issues.[7] Moreover, the range in difference between these formulations is so wide that any attempt at combining their respective messages into a single principle would seem to drain PP of almost all substantial content. This was the conclusion of Per Sandin in an excellent attempt to conceptually analyse the content of PP in the late 1990s: If we want to honour all the different formulations of PP to the full, PP must be taken to express the extremely unclear claim that in the face of some potentially dangerous activity some action should be undertaken.[8]

1.1.2 The Price of Precaution

Now, unclarities can many times be lived with. Many rules, regulations and policy declarations are unclear in their implications in various respects, but this need not bother us as long as it does not create any immediate practical problem. And if it does, the unclarity that needs to be straightened out can often be so without much need for controversy. However, in some cases, unclarities are of such a basic nature that they will create problems in almost every single instance. Furthermore, such basic unclarities almost always imply high degrees of controversy regarding how they should be eliminated or decreased. This, in turn, is due to the fact that, in order to decrease the unclarity, we need to address some issue on which people's opinions tend to conflict severely – not seldom because the issue in question awakens basic questions of ethics and values. Now, in connection to the notion of PP, there is one such central and notoriously controversial issue in particular – and, on this issue, every single suggested formulation of PP is dead silent.

Whatever unclarities are built into the notion of PP, this principle strives to say something about our (or society's, or some particular acting party's) responsibility for 'taking measures' against various potential 'threats'. No matter what this amounts to more precisely, it means that what PP prescribes or recommends (i.e. precaution, in some sense of this word) will always have a *price*. This includes, of course, straightforward monetary costs for the measures in question, but not only that.

---

[7]For example, PP has been referred to in debates about the proper treatment of animals (Bradshaw 1998), abortion (Ford 1990, Mahoney 1984) and embryo research (Ford 1990) as well as general research ethics (Kuhlau et al. 2011). According to Sunstein (2005, p. 4), the basic ideas of PP can be detected in the US responses to the threats of global terrorism. In my own country, Sweden, PP has also figured in a rather fierce public debate on the issue of the possibility of homosexual couples to adopt children.

[8]Sandin (1999). This paper also provides an informative exposition of formulations of PP available in the literature. An analysis along similar lines as Sandin's has later been presented in Manson (2002). See Bodansky (1991), and O'Riordan and Jordan (1995) for earlier observations of PP's unclarity. For comprehensive overviews of formulations of PP in different political contexts, see O'Riordan et al. (2001a) and Trouwborst (2002).

First, the precautionary measures will always in themselves create some threats or risks of their own. Secondly, they will almost always mean that some activity (assessed to be potentially dangerous enough to motivate the precautionary measure) should be avoided, postponed, modified or something similar. However, all such activities, no matter how dangerous they appear to be, will also always have some *positive* aspects (at least in the form of serving the interests of a limited number of individuals for a limited time) that will be lost or at least negatively affected by the precautionary measure.[9]

A central question in all debates on PP, therefore, regards the *proper* price of precaution. How high should it be allowed to rise – i.e., how much of increased expenditure, new threats, actual harms and lost goods should be accepted as a result of precautionary measures? And in what type of circumstances would it actually be acceptable to pay this price?

1.1.3 Precaution, Risk Analysis and Models of Rationality

This aspect connects to another difficulty in understanding PP, namely its relation to standard models of risk analysis and rational decision making routinely applied in the policy making of developed countries. These models are all variations of what is often referred to as *risk-cost-benefit analysis* – an approach to decision making derived from the default opinion in decision theory as to what determines the rationality of actions – the so-called *principle of maximising expected utility*.[10] Moreover, the standard view is that the application of these models should be backed up by as much scientifically certified information as may reasonably be required. One might say that this ideal of a rational policy making based on science embodies the contemporary version of the enlightment ideal of a society scientifically and rationally designed to the benefit of all.

As mentioned at the outset, the emergence of PP can in many ways be seen as a reaction to what is easily perceived as a failure of the enlightment ideal. And, according to some commentators of PP (supporters as well as critics), this notion should indeed be taken to mark a radical break with this ideal – also in its current form of scientifically informed application of models for risk-analysis and rational policy making. Accepting PP will thus force us to radically revise our basic conceptions of and models for the analysis of risky decisions and rational decision making

[9]Harris and Holm (2002) tries to use this fact as a general argument against PP and Sunstein (2005) seems to be of a similar bent. However, as observed in Sandin (2006), it is not the case that all versions of PP are necessarily unable to deal with these types of considerations. In fact, the theory subsequently developed in this book is explicitly constructed to prescribe a proper price of precaution.

[10]This principle as well as the standard arguments in its favour are explained and critically discussed in Chapter 3 below.

based on science.[11] It has even been claimed that the general idea of a PP should be dismissed on the basis that it is mainly an expression of irrational fear.[12]

However, in many cases, PP instead seems to be seen as more of a complement to standard models of rationality – and thus to reside well inside the boundaries of current conceptions of ideal policy making. PP is thus assumed to be quite capable of being a part of the enlightment ideal in its contemporary form, without any need for some sort of 'paradigm shift' in our basic principles of societal and scientific rationality.[13] It has even been suggested that PP may be interpreted merely as a claim for *true and consistent application* of the standard models (rather than the sort of arbitrarily constrained application that usually marks actual policy making).[14]

1.1.4 The Ideal of the Desirability of Precaution

It is far from clear, then, what PP is supposed to tell us – how it should be understood. In particular, it is extremely unclear what *practical* requirements that PP implies – both regarding how we are supposed to structure our models of policy making and what decisions should be made in particular cases. For this reason, it may come as no surprise that several commentators have argued PP to be much too unclear to be able to guide practical action, decision- or policy making.[15] If these critics are right, PP starts to look as nothing more than a mere piece of rhetoric serving as a political smokescreen for severe conflicts of interest and values. Projecting unanimity where there is an in fact grave disagreement, PP might be to the benefit of some of the parties to these disagreements. In all other aspects, however, it lacks substance and is of no practical use whatsoever.

This makes the question acute whether or not, despite the apparent force of this criticism, it is possible to find some core of shared meaning in that chaotic muddle of competing phrases that makes up our current notion of PP. I have found one single strand of hope to cling to. In 1994, Tom O'Riordan and James Cameron gave voice to the kind of criticism that has just been outlined in their conclusion of an ambitious early effort to map a wide range of prospects and problems of PP:

---

[11] See, for example, Harris and Holm (2002), Häyry (2005), McKinney (1996), McKinney and Hammer Hill (2000), Resnik (2003), Rolston III (1988) and Sunstein (2005).

[12] This is one of the main themes in Sunstein (2005).

[13] This is clear in the *Rio Declaration*, as well as other similar policy statements and political agreements, since, in direct connection to PP, explicit reference is made to a requirement of *cost-effectiveness* – a concept derived from the same body of theoretical assumptions and conceptions of rationality as the traditional models of risk analysis and rational policy making. For arguments based on the assumption that PP does not constitute a break with the ideal of scientifically informed policy making, see Buhl-Mortensen and Welin (1998), Hansson (1997), and Resnik (2003).

[14] Hansson (1999).

[15] See, for example, Bodansky (1991, 1992), Gray and Bewers (1996), Morris (2000b), O'Riordan and Cameron (1994, p. 292), Sandin et al. (2001).

> Can we honestly equate a greater commitment to precaution with a shift towards a more sustainable existence? We do not believe that the precautionary principle is either sufficiently well-defined nor properly mature as a practical concept to make such a judgement.[16]

However, after a recent (equally ambitious) revisitation of the problems involved, these leading authorities in the field of interpreting PP, joined by a third (Andrew Jordan), have forcefully claimed the possibility of finding a substantial core of PP to be quiet real. All the vagueness and ambiguity of PP to the contrary notwithstanding, the advent and continued use of the notion of PP still suggests an underlying (not yet clearly formulated) ideal about more of *precaution* being desirable in decision- and policy making on important issues. The circumstances that there are different formulations around aspiring to express this ideal and that each of these formulations may often be open to rather different interpretations do not cancel this basic fact.[17]

This is a way of perceiving the unclarities of PP to which I am highly sympathetic. In fact, this whole book is organised around the idea that there actually is such an ideal of the desirability of precaution to find and clarify. Indeed, not only that, the investigation is also built on the somewhat bolder assumption that some such clarified ideal of precaution can be shown to be acceptable and practically useful. There is a norm prescribing precaution that can both be understood (in theory as well as in its practical implications), be given a sound and valid justification and be used for solving practical issues in decision and policy making. That is my hypothesis as well as the claim I am ultimately trying to substantiate in this book.

From this perspective, the unclarities involved in PP can be seen not as a sign of emptiness, but as a natural side-effect of PP being to such a large extent the child of global political negotiation. Given the conflicts of interest between different nations when it comes to environmental and technological policy (as well as between different interest groups within these nations), it should come as no surprise that, in order to reach at least some agreement, clarity has had to be sacrificed to some extent.[18] This means, of course, that what has been agreed upon is equally unclear. However, this does not prevent the agreement from carrying at least some substance – but it is the job of policy makers, interest groups, academics and the general public to use the tools of rational and critical inquiry and debate in order to mould PP into a form where this substance becomes both intelligible and acceptable.[19]

These two requirements – intelligibility and acceptability – seems, however, to be both interdependent and potentially in conflict with each other. First, the more clear and intelligible PP is made, the higher the risk that it can be shown to lack

[16] O'Riordan and Cameron (1994, p. 292).

[17] O'Riordan et al. (2001b). A very similar conclusion seems to be reached by Sandin (2004, p. 23).

[18] Cf. the analysis made by O'Riordan and Cameron (1994, pp. 262–293).

[19] This analysis could equally well be applied to the sister notion of PP – that of 'sustainable development'. Since, in international environmental politics, PP is so strongly related to this latter notion, this of course adds to the problems involved in reaching agreement on some clearer version of PP.

acceptability. This is due to the simple fact that the acceptability of (some version of) PP is very much determined by what demands it makes in practice and that more precise versions of PP make it easier to see these demands. At the same time, however, this also means that *unclear* versions of PP become very difficult to assess from the point of view of acceptability. In order to determine whether some claim is acceptable or not we have to be able to understand what it amounts to – this, of course, goes for PP as well.[20]

On the surface of things, this may give the impression of creating an intellectual paradox: If, on the one hand, we make PP clearer we also decrease our chances of finding an acceptable version of it.[21] But, on the other, if we abstain from making it clearer we are unable to tell whether it expresses an acceptable claim or not. However, the proper conclusion to draw is, I believe, precisely that we therefore have to move the discussion on how PP is to be interpreted and understood beyond this very surface. Instead of randomly fire away suggestions for clearly formulating PP and ask ourselves whether or not they are acceptable, we need to ponder the more basic question of what can *make* any such suggested version of PP more or less acceptable.

Taking this route, it becomes necessary to direct our attention at least partly away from the peculiar mix of shallowness, pluralism and pragmatic concerns determining the treatment of PP at the level of policy making and public debate. Instead, we need to reconsider the very *basis* of PP as an ideal or recommendation for policy making – that vague ideal mentioned above of (more of) precaution being desirable in decisions regarding environmental action and the use and introduction of technology. Thus, we need to concentrate our thinking not primarily on PP itself, but on that normative or evaluative ideal that PP is supposed to express and serve to implement. What is it and what version of PP may best live up to its requirements?

This moves the inquiry beyond what at first glance may have appeared to be a question concerning the content of a concept (i.e., that of precaution) and into a field of pure normativity. What has shown itself to be at stake is not primarily what PP or the word 'precaution' *means*, but *what requirements* environmental and technological policy making *should meet* and what parameters determine to what extent any suggested requirement of this kind expresses precisely that nature and degree of precaution that is in fact *desirable*. Before we have pondered these basic normative questions, any attempt at finding an intelligible version of PP (including a clarification of the concept of precaution) that can also be shown to be *acceptable* will be futile, since we will lack the basis for assessing this very acceptability.[22]

---

[20]This is not to deny the claim of O'Riordan and Cameron (1994, p. 263) that lack of clarity seems to have been a powerful tool in reaching *acceptance* on PP. However, since this acceptance can be interpreted in as many ways as may PP, it does not imply that any specific claim regarding precaution has thereby been shown to be *acceptable*.

[21]In addition, it will probably also be more difficult to achieve wide acceptance of such a version of PP.

[22]This may explain Sandin's (2004, p. 23) sceptical conclusions regarding the prospect of finding an 'authoritative' formulation of PP. Since he does not consider what normative reasons there are

1.2 Aim, Plan and Basis

The aim of this book is to make the just outlined evaluative and normative undercurrents in the PP-debate more salient and, thereby, to facilitate a more systematic investigation of this much needed normative basis of precaution. This will be done, mainly, through a critical review of different kinds of normative arguments that may be mobilised in support of different ideals of precaution in human decision making. Ultimately, I try to defend a specific version of that ideal of the desirability precaution I above claimed to hold the key to our possibility of understanding and justifying PP.

As has been implicitly demonstrated above, PP has already provoked quite a lot of scholarly discussion on interpretation and justification. However, the question of the theoretical basis for this debate does not seem to have occupied its participants to any larger extent. Rather, out of the blue, various formulations of PP have been proposed,[23] suggestions have been made without much of support that this or that practical consequence follows from PP,[24] PP has been assumed to fit into some preconceived theoretical context, such as liberal political theory[25] or formal decision theory,[26] or the very notion of PP has, in view of the problems involved in interpreting it, been a priori declared as a misnomer we would do better without.[27]

This, I claim, is to do things backwards. If we want to find an interpretation of PP that is not only intelligible in the sense that it may support specific practical claims, but also expresses (or is based on) a *normatively valid* claim about the desirability of precaution, we need to consider how more precise ideas to this latter effect may be justified. Only then do we have at our hands the theoretical basis needed for assessing competing formulations of PP, and what actually follows from the most plausible of these. It is my hope that the investigation undertaken in this book will contribute to the construction of such a normative basis of precaution.

1.2.1 Plan of the Book

The rest of this chapter will be devoted to clearing up some preliminary conceptual issues and laying out the methodology used throughout the rest of the book. This will result in a default formulation of the vague ideal of precaution that it is the

---

to prefer one version of PP over another he simply lacks access to the type of arguments needed. However, at the same time, one particular aspect of Sandin's doubts is indeed shared also by me. See the Section 6.3 in Chapter 6 for more about this.

[23] See Sandin (1999), O'Riordan et al. (2001a), and Trouwborst (2002) for numerous examples.

[24] See, e.g., Buhl-Mortensen and Welin (1998), Francis (1996), Hansson (1999), Harris and Holm (2002), McKinney (1996), Morris (2000a), Rolston III (1988), Sunstein (2005), Thomas (1997).

[25] Klint Jensen (2002).

[26] See, e.g., Hansson (1997) and Peterson (2006).

[27] Morris (2000b).

task of the subsequent inquiry to make clear and practically useful. I will argue that this formulation – called the *requirement of precaution* – is able to capture all important issues of clarification and justification that PP may raise. Moreover, I will introduce the notion of *degrees of precaution* as a conceptual tool central for identifying unclarities that need to be straightened out, as well as for assessing the normative soundness of different interpretations of the requirement of precaution.

On the basis of these preliminary results, Chapter 2 will serve to map out several different dimensions of the requirement of precaution with reference to the above-mentioned issues of clarification and justification in terms of the degree and price of precaution. In particular, I will identify a number of parameters, the variation of which will result in different more precise versions of the requirement of precaution, each of which relevant for the assessment of clarity and justifiability. In connection to this, I will also formulate two formal conditions that any plausible version of the requirement of precaution must meet. In addition, it is also claimed that any such version must be shown to be normatively sound in the sense that it must enjoy the support of a plausible normative theory. In particular, it must be able to handle certain peculiar ethical problems regarding what actual harm it is acceptable to create in order to avoid risks of harm and it must do that in a way that is normatively acceptable. That is, a sound version of the requirement of precaution must be able to present a plausible suggestion regarding the proper *price* of precaution. Another way of putting this is that any version of the requirement must be able to justify why we should exercise exactly that degree of precaution prescribed by it.

In the following two chapters (Chapters 3 and 4), two traditional areas of normative inquiry are investigated in order to find a basis for assessing the normative soundness of different versions of the requirement of precaution. Chapter 3 investigates the possibility of supporting such versions on the basis of traditional theories of rational decision making developed in the field of normative decision theory. The conclusion of this is that in order to accomplish such a support, it is necessary to move beyond the ideals of mere instrumental rationality that define the borders of this field of inquiry and seek the support of substantial *moral* claims regarding the defensibility of creating risks of harm and our responsibility for avoiding such creation.

In Chapter 4, traditional lines of reasoning found within the philosophical field of normative ethics are therefore tried out as a possible source for justifying (a more precise version of) the requirement of precaution. After having dismissed two possible approaches (the appeal to virtue theory and the appeal to a distinction between criteria of rightness and decision procedures), the chapter is devoted to investigating whether or not it is feasible to extend the domain of application of traditional normative ethical theories of rightness and wrongness in such a way that these judge the creation of mere risks of harm (as opposed to actual harm). The outcome of this investigation is highly sceptical, no matter if the normative ethical theory taken as point of departure is deontological or consequentialist in structure. It is, however, nevertheless concluded that the investigation has reaffirmed the need for a *morality of precaution* – a normative ethical theory dealing directly with issues of creating and avoiding risks of harm.

Chapter 5, therefore, is devoted to the task of outlining the basic structure of such an ethical theory. In this chapter, many of the points made in earlier chapters resurface, but are now incorporated into the structure of a theory about, what I will call, *the moral responsibility of imposing risks*. This theory pinpoints two elements as central to the morality of creating risks: first, the balance between the seriousness of the risks involved and the actual harm created by strategies aimed at avoiding these risks; secondly, the quality of the evidence on the basis of which a decision to create a certain risk is taken. Such a theory is held to be distinct from other normative theories that may have things to say about the imposition of risks and may, consequently, give recommendations that conflict with other normative theories (such as theories of the rightness or the rationality of action). Nevertheless, I argue that the outlined theory can fill an important role in our quest for a theoretical basis for investigating the justifiability of different precise versions of the requirement of precaution. Among other things, the more precise version of the theory that I defend has the attractive potential of addressing issues of justice with regard to risks without assuming any special conception of what is just or fair.

Chapter 6, finally, is designed to substantiate the claims of Chapter 5 by giving a number of practical examples of how my theory can have substantial things to say about what risks it is responsible/irresponsible to create or allow. The examples encompass such areas as general consumerism, climate change and pollution, nuclear power, information technology and biotechnology. Among other things, these applications illustrate that a sound morality of precaution have important implications for questions about global distributive justice, but also actualise a number of problems regarding the structures for policy making both within nations and in international politics. On the basis of this, I sketch a model of precautionary policy making within the context of single nations and then concludes by addressing a number of overarching issues regarding the relation between the prescriptions of the ideal of precaution and the ideals of the enlightment and liberal democratic societies. I suggest that the morality of precaution makes a case for more democratic and robust political structures on the global level and that this implication can be harmonised with both liberal democratic political ideals and the ideals of the enlightment.

1.2.2 The Requirement of Precaution

As has been repeated in variations above, there is an abundance of suggested formulations of PP to be had and each of these can be understood in a multitude of different ways. Nevertheless, the point of departure for the investigation undertaken in this book is that all these formulations have something in common – they all aim to express, at least indirectly, a normative ideal regarding the desirability of precaution. Since the project of my investigation is to clarify this unclear idea, it is necessary to find a formulation of it that can serve as a point of departure for the discussion to follow. Moreover, I stated above that this idea is something beyond those formulations of PP that can be found in, e.g., policy and legal documents, public debates etc. and that the search for it therefore necessitates us to focus our attention away from this surface reality of PP. However, at the same time, that ideal of the desirability of precaution that PP is here assumed to express obviously cannot be formulated completely without reference to the PP's of the real world. If nothing else, there must remain a connection of this kind that directs the discussion of the underlying ideal of precaution in such a way that what is being said about it can eventually be applied to those formulations of PP that are actually around and in practical use.

Now, in spite of the multitude of available formulations of PP, if we look at them a little bit closer, the following three underlying thoughts are recurring[28]:

The Requirement of Precaution:
Activities, which may bring great harm, should not be (or be allowed to be) undertaken unless they have been shown not to impose too serious risks.

The Proof Requirement of Justifiable Policy Claim:
Policy measures against some activity that may bring great harm may be justified even if there is no scientific proof that this activity imposes (or would impose) this harm.

The Burden of Proof Requirement:
Showing that some condition for the permissibility of activities is met is the responsibility of those who propose to undertake the activity in question.

All of these ideas are clearly normative,[29] however only the first implies by itself any clear recommendation of *precaution*.

The *proof requirement of justifiable policy claim* (which, as we have seen, expresses the basic thought of the PP of the *Rio Declaration* as well as several other influential policy documents[30]) issues a claim about the *possibility* of justifying precautionary *policy measures* in the light of a lack of scientific evidence. However, it does not state when such measures *are*, in fact, justified (and what would make them so).[31] It is also silent regarding what precautionary policy measures may or may not be justified when scientific evidence *is* in fact available. And, perhaps most obviously, since precautionary policy measures may themselves be sources of harms and risks that cannot be scientifically proven, *the proof requirement of justifiable policy claim* is far from being as clear, plausible and practically applicable as some commentators have hinted.[32] Neither does this claim suggest any reason for its specific component regarding the possibility of justifying *policy measures*

---

[28]Cf. Sunstein (2005, pp. 18–20), among others.

[29]Oddly enough, a few critics of PP have perceived this rather obvious fact to be a serious problem. See Sandin et al. (2001) for a clear presentation as well as convincing dismissal of this argument.

[30]Sandin (1999), O'Riordan et al. (2001a), and Trouwborst (2002).

[31]Sandin (2004, pp. 15–16) makes a similar observation in terms of a distinction between "argumentative" and "prescriptive" versions of PP, where the former denotes ideas in line with the *proof requirement of justifiable policy claim.*

[32]For example, Bradshaw (1998), Francis (1996), and Haller (2000).

in particular (why not also/only precautionary actions that are not to be counted as expressions of public policy?). In all these ways, it begs for a justificational basis.[33]

The *burden of proof requirement* is an administrative recommendation to be found in a variety of contexts where some version of PP is suggested.[34] However, as such, it is quite applicable to any proposed requirement that our actions are supposed to meet – be it in general or according to some policy. As an extreme illustration, suppose that environmental legislation was to request nothing more of potential polluters than that they supply some vague reason for believing their business to be economically profitable the coming few years. In this case, the burden of proof requirement will still have been observed to the fullest. In itself, therefore, this requirement seems to have nothing to do with the idea of the desirability of *precaution*.[35] It also shares the need of the *proof requirement of justifiable policy claim* to acquire some form of basis for its own justification.

It is, furthermore, unclear what acting parties are addressed by these norms. They could, of course, be taken as formulations proposed for inclusion into the legal system of a country (or in international law). However, that would just shift the issue to *why* they should be so included. Another interpretation is to see them as underlying normative claims, directed at individuals in general and capable of explaining why some version of PP should be included in, e.g., environmental legislation. Such norms may be either basic or intermediary between basic normative principles and the justification of specific regulative political measures.

The conception of statements to be included in actual regulation seems to me to be well suited for the *burden of proof requirement*, since this idea seems so clearly instrumental in nature. The idea of placing the burden of proof on those who want to do something that is permissible only under certain conditions is surely not a proposal that this is of value as such. Rather, the idea is that the adoption of such a rule in some cases will be beneficial for the realisation of some ideal, e.g., that people do not behave recklessly. However, since this requirement does not really say anything about what being cautious or exercising precaution amounts to or requires, with the exception of some brief remarks in the final chapter, it will henceforth be ignored.

---

[33]As demonstrated by Klint Jensen (2002), the primary pull of this particular aspect of PP seems to be that it is easily accommodated into a liberal democratic political scheme. This, however, to the price of abstaining from any specific recommendation regarding the taking of measures, and, therefore, of passing over in silence all of the unclarities regarding the formulation of a more precise version of PP capable of guiding action as well as decision- and policy making (cf. Morris 2000b, pp. 13–15). After all, in the liberal context, the formulation does not really say anything more than suggesting that the proof-standards of science need not be the proof-standards of liberal politics.

[34]Sandin (1999), O'Riordan et al. (2001a), and Trouwborst (2002).

[35]In spite of this, it is my own personal experience that many people – not least environmental activists and those with similar stakes in the debate on how to understand PP – actually believe this claim for a reversed burden of proof to express the very essence of what PP has to say.

The *requirement of precaution*, in contrast, seems very fitting indeed for being seen as an underlying norm, capable of underpinning more specific norms and regulations regarding various acting parties and types of action. For example, it may be used to justify the *burden of proof requirement*. And, in my view, it also seems a very fitting candidate indeed for being precisely that justificational basis so badly needed by the *proof requirement of justifiable policy claim*. The *requirement of precaution* is both more general (speaking not only about the justification of policy measures when scientific evidence is lacking) and more specific in its normative message (saying more than merely stating the possibility of justifying some actions under some conditions). In both these aspects, it by far supersedes the *proof requirement of justifiable policy claim* as a fitting point of departure for the kind of investigation undertaken in this book.

In this rather harsh way, then, I am dismissing both the *burden of proof requirement* and the *proof requirement of justifiable policy claim* as plausible candidates for some deeper ideal of the desirability of precaution that may serve to clarify PP. However, I do want to point to a common feature of them both that will still remain an important aspect of the discussion to follow. This feature is: they both serve to highlight the intended use of PP to be *directly political* (directly in the sense that it actually guides the substantial content of policy, not only serves as a piece of rhetoric used for influencing public opinion). This means that the normative basis of PP pursued in this book must be so designed that it can justify some more precise version of PP that may serve this purpose. As will be seen later on, this aspect of PP will have substantial bearing on the more theoretical discussion to follow, but also create some problems when it comes to the practical application of the results of this discussion.

To be true, the requirement of precaution is unclear in several respects, but that is an advantage in the present context. What we need as a starting point in our pursuit of the normative basis of PP is not a ready-made precise opinion. On the contrary, what makes the requirement of precaution a fruitful tool for the investigation to follow is that it is a formulation that may plausibly be taken to capture those *unclear* ideas about the desirability of (more) precaution referred to above. In this respect, its unclarity is an outstanding virtue from a methodological point of view.

In the rest of this book, therefore, I will use the requirement of precaution, understood as a proposed basic or intermediary underlying norm regarding the desirability of precaution, as starting point for a discussion of its proper rationale. On what basis might it be justified and what more precise interpretation of it can be supported from this justificational basis?

1.2.3 Degrees of Precaution

In the next chapter, the task will be to map out the various sources of the unclarity of the requirement of precaution that need to be addressed in the discussion to follow. However, unclarity is a global and limitless linguistic disease – it affects all notions and phrases and there seems to be no limit to its occurrence. No matter how clear we make some concept or statement, it will always be possible to find some further vagueness or ambiguity. This means that any investigation of the unclarities of some concept or notion has to concentrate on those that have a special bearing on some other issue where this concept or notion has a part to play.

In the case of the notion of PP and the concept of precaution such 'other issues' comprise a wide range of subjects stretching across a smorgasbord of academic disciplines. Economics, law, linguistics, political science and sociology are but a few of those fields where issues referring to PP and precaution enter the scene. However, given the differences between these disciplines as to what is their centre of attention, it can hardly be expected that the same dimensions of unclarity will be of equal interest in all settings. On the contrary, we should expect that, in each field, some specific types of unclarity will be of special interest. A project of mapping out the unclarities of the concept of precaution therefore needs to be tailor-made for one specific aim. That is, the choice of 'parameters' along which the unclarities of the requirement of precaution are assessed must be firmly guided by one specific 'other issue' to which the notion of PP and the concept of precaution are to be applied.

One such other issue of interest from the point of view of both the analysis of unclear concepts in general and the requirement of precaution in particular is that of investigating the level of *precision* of various versions of PP. This is what seems to have originally guided Per Sandin in his seminal conceptual analysis of PP referred to earlier. In trying to find a formulation of PP that is specific enough to be called PP at all, but flexible enough to be compatible with all known (and mutually incompatible) formulations of PP, Sandin has been sketching the methodological starting point for a semantic project aiming at comparing the level of precision of competing versions of PP.[36] In later writings, Sandin has to some extent modified the original project.[37] However, he has also made efforts to outline a model for the practical operationalisation of PP where the hunt for precision continues to play a dominant role.[38]

Now, since the standard complaint about PP concerns its lack of clarity, designing a model aiming at hunting down versions of PP with higher levels of precision may appear as exactly the thing to do. However, from the point of view of the 'other issue' that forms *my* point of departure – that of the morality and normative basis of precaution – this alleged adequacy becomes a bit dubious. Pondering what justificational basis competing versions of the requirement of precaution should be assessed from and what such version is to be preferred from this perspective, assessing and

[36] Sandin (1999). Sandin uses the term "strength" and equates this with "degree of precaution" (p. 898). However, he also states that the strength of a version of PP determines its precision (p. 898). Moreover, it is clear from the construction of his theoretical setup that what he primarily wants to be able to do is to compare semantic precision. What he calls "degree of precaution" should therefore not be confused with my own use of this notion (see below).

[37] In particular, he has added the idea of comparing how "extensive" and "demanding" different versions of PP may be (Sandin 2004, p. 16).

[38] Sandin and Hansson (2002).

comparing levels of precision may, of course, come into play. However, there is no reason to *assume* precision to have the overwhelming importance designated to it in the context where Sandin performed his analyses. For, after all, the most basic issue regarding PP (as with any norm or rule) is whether or not we should comply to its prescriptions – i.e., to what extent it can be justified or shown to be acceptable. And, since it is not given what level of precision we will find in that version of PP (or the requirement of precaution) that can actually be so justified, whether or not increased precision of a version of PP or the requirement of precaution is to be seen as a virtue or not is a completely open question.[39]

What, then, *is* important to know about the unclarities of PP, the concept of precaution and – in particular – the requirement of precaution from the point of view of the pursuit of a normative basis of PP? Here is a simple answer: in assessing and comparing competing versions of the requirement of precaution we need to pay attention to precisely those unclarities that may have a bearing on the normative soundness or acceptability of these versions. Now, what determines this soundness or acceptability will depend on what underlying theory we apply in order to accomplish such a normative justification. In the end, therefore, what is of importance is to map out unclarities corresponding to the choice of the underlying theory of normative justification.

As mentioned above, Chapters 3, 4 and 5 will be devoted to critically assessing competing theories of this kind. Already now, however, I will claim that there is one type of conceptual consideration that has a supreme standing in that discussion, namely what I will henceforth refer to as the *degree of precaution* prescribed by a version of the requirement of precaution. This notion expresses the basic idea that precaution is not a binary concept, but a continuous one: *all precaution is on a scale of more or less*. The more difficult, costly and demanding it is to have the conditions laid out by a certain version of the requirement of precaution met, the greater the degree of precaution it prescribes.[40] This difficulty, costliness and demandingness can in turn be explained in terms of three main factors: First, the time and resources needed for meeting the requirement. Second, destruction of values or creation of risks of such resulting from meeting the requirement. Third, values that might have been gained by pursuing other activities rather than meeting the requirement and which are therefore lost due to it being observed (i.e. the alternative costs of meeting the requirement of precaution).

---

[39]I am thus rejecting as evident assumptions the requirements of completeness, determinism, predictability and uniqueness for the operationalisation of PP held out by Sandin and Hansson (2002, pp. 465–466). Whether or not (and to what extent) such factors should be given weight in the assessment of competing versions of PP depends on what weight they can be argued to have from the normative basis of precaution pursued in this book.

[40]For some readers, this notion may be confusing, since it may be understood in terms of how small threats that must be in place in order for the requirement of precaution to prescribe us to take precautionary measures. According to this understanding, the smaller or less serious such threats need to be, the more cautious we will be if abiding by these prescriptions and, thus, the more of precaution is prescribed by them. As will be seen in Chapter 2, this factor does come into play as one of several others influencing the degree of precaution (in my sense).

As may already have been noticed, this explanation of what determines the degree of precaution fits neatly to the notion of a *proper price* of precaution mentioned earlier. The higher degree of precaution prescribed by a version of the requirement of precaution, the higher the price of precaution it claims to be acceptable – and vice versa. Thus, the concept of degrees of precaution connects tightly to that issue which, at the outset of this chapter, I argued to hold the key to the whole debate on PP: the question of what price it is acceptable to pay in order to achieve precaution. And, as we will see in later chapters, it is primarily on this issue that rivalling theories about the normative basis of PP both disagree and run into various types of problems.

In view of this, those dimensions of the unclarity of the requirement of precaution mapped out in the next chapter are those that I have found to be important determinants of the degree (and hence price) of precaution. The central importance of this latter factor in the pursuit of a normative basis of precaution will be demonstrated by the actual normative discussion in Chapters 3, 4 and 5. Until then, for those readers who have not been convinced by what has been said so far, I beg for some tentative (albeit cautious!) trust in my judgement on this point.

References

Bodansky D. "Scientific Uncertainty and the Precautionary Principle." *Environment* 33 (7) (1991): 43–4.

Bodansky D. "Commentary: The Precautionary Principle." *Environment* 34 (3) (1992): 4–5.

Bradshaw R.H. "Consciousness in Non-human Animals: Adopting the Precautionary Principle." *Journal of Consciousness Studies* 5 (1) (1998): 108–14.

Buhl-Mortensen L. and S. Welin. "The Ethics of Doing Policy Relevant Science: The Precautionary Principle and the Significance of Non-significant Results." *Science and Engineering Ethics* 4 (4) (1998): 401–12.

Ford N.M. "Ethics, Science and Embryos." *The Tablet*, February 3 (1990): 141–2.

Francis J.M. "Nature Conservation and the Precautionary Principle." *Environmental Values* 5 (3) (1996): 257–64.

Graham J.D. "Perspectives on the Precautionary Principle." *Human and Ecological Risk Assessment* 6 (3) (2000): 383–5.

Gray J.S. and M. Bewers. "Towards a Scientific Definition of the Precautionary Principle." *Marine Pollution Bulletin* 32 (11) (1996): 768–71.

Haller S. "A Prudential Argument for Precaution Under Uncertainty and High Risk." *Ethics and the Environment* 5 (2) (2000): 175–89.

Hansson S.O. "The Limits of Precaution." *Foundations of Science* 2 (1997): 293–306.

Hansson S.O. "Adjusting Scientific Practice to the Precautionary Principle." *Human and Ecological Risk Assessment* 5 (5) (1999): 909–21.

Harris J. and S. Holm. "Extending Human Lifespan and the Precautionary Paradox." *Journal of Medicine and Philosophy* 27 (3) (2002): 355–68.

Häyry M. "Precaution and Solidarity." *Cambridge Quarterly of Health Care Ethics* 14 (2005): 199–206.

Klint Jensen K. "The Moral Foundation of the Precautionary Principle." *Journal of Agricultural and Environmental Ethics* 15 (1) (2002): 39–55.

Kuhlau F., A.T. Höglund, K. Evers, and S. Eriksson. "A Precautionary Principle for Dual Use Research in the Life Sciences." *Bioethics* 25 (1) (2011): 1–8.

Mahoney J. *Bioethics and Belief*. London: Sheed & Ward, 1984.

Manson N. "Formulating the Precautionary Principle." *Environmental Ethics* 24 (2002): 263–74.
McKinney W.J. "Prediction and Rolston's Environmental Ethics: Lessons from the Philosophy of Science." *Science and Engineering Ethics* 2 (4) (1996): 429–40.
McKinney W.J. and H. Hammer Hill "Of Sustainability and Precaution: The Logical, Epistemological, and Moral Problems of the Precautionary Principle and Their Implications for Sustainable Development." *Ethics and the Environment* 5 (1) (2000): 77–87.
Morris J. ed. *Rethinking Risk and the Precautionary Principle*. Oxford: Butterworth-Heinemann, 2000a.
Morris J. "Defining the Precautionary Principle." In *Rethinking Risk and the Precautionary Principle*, edited by J. Morris. Oxford: Butterworth-Heinemann, 2000b.
O'Riordan T. and J. Cameron, eds. *Interpreting the Precautionary Principle*. London: Earthscan Publications, 1994.
O'Riordan T. and A. Jordan. "The Precautionary Principle in Contemporary Environmental Politics." *Environmental Values* 4 (3) (1995): 191–212.
O'Riordan T., A. Jordan, and J. Cameron, eds. *Reinterpreting the Precautionary Principle*. London: Cameron May, 2001a.
O'Riordan T., A. Jordan, and J. Cameron, eds. "Reinterpreting the Interpretation." In *Reinterpreting the Precautionary Principle*, edited by T. O'Riordan, A. Jordan, and J. Cameron. London: Cameron May, 2001b.
Peterson M. "The Precautionary Principle Is Incoherent." *Risk Analysis* 26 (3) (2006): 595–601.
Resnik D.B. "Is the Precautionary Principle Unscientific." *Studies in the History and Philosophy of Biological and Biomedical Sciences* 34 (2003): 329–44.
Rolston III. H. *Environmental Ethics*. Philadelphia: Temple University Press, 1988.
Sandin P. "Dimensions of the Precautionary Principle." *Human and Ecological Risk Assessment* 5 (5) (1999): 889–907.
Sandin P. *Better Safe than Sorry: Applying Philosophical Methods to the Debate on Risk and the Precautionary Principle*. Stockholm: Royal Institute of Technology, 2004.
Sandin P. "A Paradox out of Context: Harris and Holm on the Precautionary Principle." *Cambridge Quarterly of Health Care Ethics* 15 (2) (2006): 175–183.
Sandin P. and S.-O. Hansson. "The Default Value Approach to the Precautionary Principle." *Human and Ecological Risk Assessment* 8 (3) (2002): 463–71.
Sandin P., M. Peterson, S.O. Hansson, C. Rudén, and A. Juthe. "Five Charges Against the Precautionary Principle." *Journal of Risk Research* 5 (4) (2001): 287–99.
Sunstein C.R. *Laws of Fear: Beyond the Precautionary Principle*. Cambridge: Cambridge University Press, 2005.
Thomas V.G. "Attitudes and Issues Preventing Bans on Toxic Lead Shot and Sinkers in North America and Europe." *Environmental Values* 6 (2) (1997): 185–99.
Trouwborst A. *Evolution and Status of the Precautionary Principle in International Law*. The Hague: Kluwer Law International, 2002.

Chapter 2
Dimensions of Precaution

How is the following claim to be understood and justified? What theoretical presuppositions does it make use of? What are its practical implications? What more precise version of PP can it support?

> *The Requirement of Precaution*:
> Activities, which may bring great harm, should not be (or be allowed to be) undertaken unless they have been shown not to impose too serious risks.

The task of this chapter is to supply a foundation for the further investigation of these issues. Moreover, in light of what was said at the end of the foregoing chapter, the content of this foundation should be an outline of those unclarities that are of consequence for what degree of precaution is prescribed by the requirement.

There are six aspects of the requirement of precaution which seem to contribute to its lack of clarity in such a way:

1. It does not specify its assumed *theory of value*,
2. it considers only actions which *may bring great harm*,
3. it requires something to be *shown*,
4. it speaks about *risks*,
5. it assumes that risks may be *too serious*, and
6. it does not specify its own *level of application*.

Each of these aspects may be seen as a dimension, where variations affect the degree of precaution prescribed by the requirement. However, the function of these dimensions for the requirement, i.e. *how* they influence its more substantial content in this respect, differ from each other.

In the following, I will first say something about dimensions 1 and 6 and thereafter, for reasons to be explained, leave these aside for the time being. These aspects will reappear in the attempt to *apply* the theory on the ethics of risk to practical cases undertaken in the final chapter.

The rest of this chapter will attempt to explain the differing functions of dimensions 2–5 in that order. Apart from supplying a conceptual foundation to build on in

C. Munthe, *The Price of Precaution and the Ethics of Risk*, The International Library of Ethics, Law and Technology 6, DOI 10.1007/978-94-007-1330-7_2,

later chapters and clearing up a number of more or less common confusions under way, this work will also result in three substantial claims.

First, no single one of these dimensions can be seen as more fundamental than any other in relation to how they affect the degree of precaution prescribed by the requirement of precaution. On the contrary, any change of this degree that results from a change in one of the dimensions may be counteracted by a change in some other dimension. In this sense, the degree of precaution prescribed by the requirement is truly holistically determined – a conclusion of importance both for increasing our understanding of why the debate on PP demonstrates so much confusion, and for reinforcing the reasons for the direct normative approach to the question of the interpretation of PP taken in this book.

Secondly, close inspection of the dimensions of precaution will result in two formal conditions that any acceptable version of the requirement of precaution needs to meet. This is of great importance for several reasons. First, it saves us from the tedious process of demonstrating for each member of a large family of hopelessly extreme and/or single-minded interpretations that they are in fact as faulty as they appear at first inspection. However, it also serves to let PP off the hook of an equally large, extreme and/or single-minded family of criticism!

Third, the investigation of these dimensions will serve to clarify further the fundamental role of the issue of the proper price of precaution. Moreover, this will demonstrate clearly what is in fact at stake when pondering from what normative basis the requirement should be justified and clarified.

2.1 Values, Levels and Time-Horizons

2.1.1 Values

The very essence of the requirement of precaution is to say something about when it may be acceptable to risk the destruction of *what is of value* (and when it may not). Thus, one obvious unclarity of the requirement is that it lacks the information of a specified theory about what makes states of affairs more or less valuable and how the values of such states may be compared and balanced against one another.

First and foremost, this theory is needed for the clarification of dimension 2. We need to specify what makes for a harm and what determines the magnitude and seriousness of harms. However, it is also needed for the clarification of what below will be set forth as the most plausible scheme for conceiving of the magnitude and seriousness of risks (dimensions 4 and 5), since this scheme involves taking into account how bad it would be for risks to be actualised.

The political context from which PP has emerged (and thus the source of the normative ideal that the requirement of precaution is taken to express) has seldom been very clear on questions such as these. The *Rio Declaration* and other international agreements on environmental policy often hint at a rather unspecified form of (axiological) *anthropocentrism* – i.e. the view that the value of a state of affairs is determined solely by how human beings (and, perhaps, equivalent creatures) fare in some respect. The question "in what respect, more precisely?" actualises a traditional lot of competing theories of value – phrased in terms of well-being, the good life, quality of life etc. – discussed in normative ethics and moral philosophy since the very beginning of these subjects.[1] In addition to this, there are competing ideas on the proper aggregation and balancing of valuable states, some of which will be traceable in some of the discussions of this book, but that for the most part will be left open.[2]

However, an increasing trend in the web of issues, problems and conflicts surrounding PP is that the situation of humans is no longer seen as the only determinant of value. This is most obvious in the case of animal protection acts in many countries around the world. How we treat and influence the lives of (large and clearly sentient) animals is to an increasing extent being valued in its own right (and not only in terms of how this affects human beings). However, even non-sentient and non-living entities in the form of biological systems – such as species, landscapes, ecosystems etc. – are in a similar manner beginning to be accorded values of their own. Not only is this an increasingly popular view among environmental activists and 'green' politicians, but also it is in fact expressed (at least implicitly) in some environmental policy documents and even in actual legislation.[3]

Related to my quest for a clarification of the requirement of precaution, what this means is that pondering only traditional disagreements and arguments in ethics on the question of value is not sufficient. The context of PP makes it necessary to consider also a heap of more novel suggestions according to which nature (or some part of nature) harbours or is itself of value quite independently of how human beings or other sentient creatures fare.[4] While the practical consequence of taking a stand in the traditional discussion on the *human* good may often be doubted, including these radical thoughts among the ideas that might clarify, for example, what may count as harm or a risk according to the requirement of precaution, makes this choice into a question of the highest practical priority. It is suddenly very probable indeed that the choice of value theory will have tremendous effect on the practical consequences of the requirement – not least what degree of precaution it prescribes.[5] In effect, just

---

[1] See, for example, Parfit (1984, pp. 3–4), and Brülde (1998).

[2] This concerns, for example, a large part of the very problematic issue of how possible future bearers of value (e.g., people with decent qualities of life that might come to exist in the future) should be taken into account. See Parfit (1984, part 4), and Arrhenius (2000), for detailed and provocative treatments of this issue.

[3] For example, the Swedish Environmental Protection Act ("miljöbalken") makes a salient distinction between considerations of human welfare, animal well-being and environmental values. Although one should perhaps not make too much of this, similar more or less clearly stated distinctions can be found in policy documents and laws in many countries around the world.

[4] See, for example, Andersson (2007).

[5] Cf. the arguments pursued in Stenmark (2002) regarding the practical relevance of theories of environmental ethics, and Allhoff (2009) regarding the impact of this relevance on the choice of interpretation of PP.

as the vague ideal of more of precaution being desirable in societal decision making may be taken to express the complaint that sound models of good policy making are not consistently applied in practice, it may also be interpreted as a complaint regarding the base of values used in such applications being too narrowly defined.

In spite of this, however, I will in the following neglect to discuss this dimension of precaution any further. Important as it may be for the practical application of the requirement of precaution, the issues it actualises can be discussed quite independently. Although being of consequence for what demands will be made by a fully specified version of the requirement, these issues are thus not necessarily connected to judgements about the acceptability of risk impositions. And, since issues about the clarification and justification of such judgements lie at the very heart of the requirement of precaution, I will in the following concentrate exclusively on these.

2.1.2 Levels and Time-Horizons

This takes us to the dimension 6, regarding the *level of application* of the requirement of precaution. Unlike the dimension of value theory, this one does not really concern the substantial content of the requirement. Rather, it concerns to whom and what the requirement of precaution should be taken to *address itself*. The unclarity hence is about what problems or questions that the requirement more precisely purports to solve or answer. And this, of course, can be discussed quite independently of what actual claims are made by the requirement about the acceptability of risk impositions, our reasons for taking measures to prevent such impositions, et cetera.

This dimension of precaution captures what decision theorists often call the *framing* of decision problems – i.e. according to what criteria different such problems are distinguished from one another. For example, how the particular question to which the requirement is supposed to provide answers is specified with respect to *what acting parties are addressed*[6] may produce quite peculiar effects. This, in turn, is largely due to the phenomenon of what is sometimes called coordination problems – i.e. the fact that even perfect observation of a decision rule on one level of acting parties may result in the complete opposite result at another level.

As an illustration, Sven-Ove Hansson has remarked in direct connection to PP that even if it were to be successfully applied *locally* by every single nation, this would not guarantee that the recommendations posed by PP (whatever they are) would be met on a *global* scale.[7] For, as has been repeatedly demonstrated in both decision theory and moral philosophy, even if *each of us* acts according to some norm, this is no guarantee for the aggregated result of our actions to comply with this

[6] Individual people? Organised groups? Public institutions? Commercial companies? Political regions? Countries? Quasi federative associations? Multinational organisations? All of these? Some of these? Just one?

[7] Hansson (1997).

norm.[8] For example, each nation in the world successfully applying the requirement of precaution may still add up to a result where the world at large is exposed to activities that may cause great harm and that have not been shown not to bring too serious risks.[9]

As if this complication was not enough, it seems that the very same phenomenon may occur not only because of the partition of different decisions in space (such as each nation deciding its own policy). Also the partition of the decisions taken by one single party *over time* may have similar effects. For example, a temporal series of decisions, each conforming to the requirement of precaution, may add up to an end result where activities that may cause great harm and that have never been shown not to bring too serious risks are still allowed. For this reason, not only the level of application of the requirement needs to be specified, but also its *time-horizon* – i.e. how it distinguishes between different situations or decision problems over time.

Although these phenomena may occur no matter what norm or decision rule we consider, in the case of the requirement of precaution, the situation is especially complicated, since it contains the element of the 'may bring great harm' clause that has to be met in order for the requirement to be applicable at all. No matter how this clause is specified, it will allow for series or lumps of activities each of which not falling under the clause but together making up a joint activity falling under the clause.

The general problem, then, seems to be how to interpret the requirement of precaution in the light of these aspects. How should the problem or problems that the requirement is supposed to solve be formulated with respect to the partition of acting parties and the options faced by these (in space as well as time)? Since these questions do not directly concern the substantial content of the requirement (they can be posed in relation to any norm), they may be thought not to be among those unclarities used for working out a theory of the normative basis of precaution. However, as will be explained in Chapter 5, the fact that many of the risk scenarios explaining the perceived need for PP and an underlying ideal of precaution involve the sort of coordination problems illustrated by the dimensions of levels and time-horizons, seems to imply that certain ways of constructing a more precise theory on the morality of precaution are in fact excluded. Moreover, the questions about levels and time-horizons are of tremendous importance for the application and practicality of the resulting version of the requirement of precaution. I will, therefore, return to the issues about the level and time-horizon of the requirement also in the final chapter.

---

[8]In decision theory, the paradigmatic example of this phenomenon is, of course, the so-called prisoners' dilemma. Regarding moral philosophy, several parallel cases have been described by Parfit (1984, part 1).

[9]The same phenomenon is in fact true of any norm or decision rule formulated by decision theorists (such as the principle of maximising expected utility, the maximin principle etc.) and moral philosophers (such as utilitarianism, Kantian deontology, theories of rights etc.).

2.2 May Bring Great Harm

With these preliminaries set aside, let us now turn to those dimensions of precaution that have direct bearing on the forthcoming construction of a normative basis of PP. The first of these dimensions is expressed by the limit set by the requirement that it speaks only of activities that 'may bring great harm'. Either an activity may bring great harm, in which case the requirement is applicable, or it may not, in which case it is of no concern of the requirement. It should be obvious, therefore, that variations in this dimension may have a tremendous impact on the degree of precaution prescribed by the requirement.

2.2.1 De Minimis Risk and the Need for a Limit

The inclusion of this 'self-limitation' is inspired by the suggestion that PP has to employ some limit on how unlikely a risk-scenario of an activity can be in order for this activity to be within PP's range of applicability. Otherwise, even the most trivial activity would have to be subjected to the requirements of PP, since everything we do *might* (although this is often extremely improbable) produce some kind of undesirable outcome.[10]

Among decision theorists and risk-analysts this idea is known as '*de minimis* risk' (i.e. risks that may be overlooked due to their actualisation being sufficiently unlikely) and has been the subject of various debate and criticism.[11] In my view, the most fundamental objection to this idea is that it gives way too much weight to the probability aspect of risks. In standard models of risk-analysis, the gravity of a risk is determined by an even balancing of its probability aspect (i.e. how likely it is that the risk is actualised) and its *value* aspect – how undesirable it would be if the risk were to be actualised. However, when the *de minimis* idea is added, even the smallest difference in probability suddenly receives a tremendous importance in its own right – as long as this difference is able to tip the scale regarding whether the risk in question is *de minimis* or not.[12]

In the context of PP this objection seems to be especially forceful. As was seen in Chapter 1, it is obvious that PP has emerged as an answer to our fears of *especially undesirable* events resulting from human activities. In effect, also the fact that a possible negative outcome of some activity is not undesirable enough would seem to be a factor relevant for whether or not this possibility may be overlooked or not. For these reasons, I have preferred the broader idea of a limit for the requirement's range

---

[10]Hansson (1997), and Sandin et al. (2001).

[11]Whipple (1987). See Peterson (2002) for an accessible overview of various suggestions as to how the *de minimis* idea should be operationalised in actual policy making, as well as criticism of these suggestions.

[12]See Sandin (2005) for criticism of this type in relation to the particular idea that the *de minimis* likelihood levels are to be specified on the basis of 'natural' risk levels.

of applicability in terms of both the likelihoods and the values at stake, with some emphasis on the latter. In order for the requirement of precaution to be applicable to an activity, this activity must be such that it *may* produce a result, which in turn is a *great harm*.

This notion is unclear in two ways, first, in what it means that an action *may* bring something about and, secondly, what is to count as great (rather than small or modest) harm. Both these sources of unclarity connect to the discussion on *de minimis* risk mentioned above. In the following, I will therefore relate them to the standard argument put forward in support of this idea.

2.2.2 The Argument from Decision Costs

The following seems to be a general truth about the dimension of precaution presently under consideration: the less that is required of a scenario of great harm for it to be a scenario that *may* occur, and the lower the amount of harm needed to make up a *great* harm, the wider the range of applicability of the requirement of precaution. As a rule, such a wider range will also lead to increased difficulties of meeting the conditions of the requirement, since more options in each situation of choice – as well as more possible outcomes of these options – will be subject to them. For example, if 'may' is interpreted so weak as to mean roughly 'not logically incompatible with currently known scientific data', in most situation, the requirement will be equally applicable to all alternatives open to an agent, including the alternative of 'doing nothing'. And if the criterion of 'great harm' requires very little harmfulness (such as someone suffering a brief and weak unpleasantness of some kind), the same effect will yield even if 'may' is much more strongly interpreted.

The standard argument in favour of the *de minimis* idea is that unless we set a limit to what we have to take into consideration when making decisions, the costs for making these decisions will become unacceptably high.[13] Although I argued above that the particular idea of *de minimis* risk should be abandoned, it seems to me that this argument from decision costs can be applied to the 'may bring great harm' dimension of precaution. As has just been mentioned, variations in this dimension affect how inclusive the requirement's range of applicability will be. And the more inclusive it is, the more difficult it will be to meet the conditions set by the requirement. Now, what this means in practice is that greater inclusiveness will result in increased decision costs – more resources and time will have to be spent on meeting the conditions of the requirement than if its range of applicability had been less inclusive.

In the context of the *de minimis* debate, the general view seems to be that what is to count as unacceptably high decision costs is determined by standard rules of

[13] Mumpower (1986), Shrader-Frechette (1985), Weinberg (1985).

rational decision making, such as the principle of maximising expected utility.[14] However, since in the debate on PP it is highly unclear how this principle relates to the idea of the desirability of precaution, this view of what is to count as unacceptably high decision costs cannot be taken for granted in the present context. Rather, from the point of view of the requirement of precaution, what is to count as decision costs and what determines whether or not these are too high in a particular case, would have to be determined by considerations regarding whether or not paying these costs goes against the spirit of the requirement – i.e. the more general idea of the desirability of precaution expressed by it.

So modified, it seems that the argument from decision costs can be directly related to the issue of what *degree of precaution* is prescribed by the requirement. Simply put, the higher the decision costs resulting from a specified version of this requirement, the higher the degree of precaution it prescribes. Moreover, since the degree of precaution prescribed by the requirement also determines what *price* of precaution it tells us to pay, the questions of what is to count as decision costs and what determines whether or not these are too high can be reduced to the issue of what is the *proper* price of precaution. In other words, the question of how the 'may bring great harm' part of the requirement of precaution should be specified must be decided on the basis of an answer to this more basic normative issue.[15]

As has already been briefly noted and as will become more and more salient as we go along, what this answer would consist in is no clearer than the general ideal of the desirability of precaution. We might, therefore, seem to be caught in an intellectual paradox where, in order to explain the requirement of precaution, we have to make clear the limit for unacceptably high decision costs, but at the same time, in order to do this, we have to have access to a clear idea of the requirement of precaution (specifying what is the proper price of precaution). I will return to this problem in Chapter 3 and there argue that, in order to have a chance of resolving this issue, the discussion needs to take into consideration moral ideas, which may help clarify the basic notion of the desirability of precaution.

However, before that, it is now time to turn to the other dimensions of precaution. Doing so, we will see that the connection between the idea of a limit for the range of applicability of decision rules and norms (such as dimension 2 of the requirement) and the argument from decision costs is rather more complicated than what has been set out above. At the very least, given the context of the requirement of precaution and the modification of the argument from decision costs made above, this

---

[14]If the consideration of scenarios with a likelihood below a certain degree would make the whole process of deciding what option maximises expected utility suboptimal (from the point of view of this very same principle) compared to acting on chance or some other principle, the decision costs of including these scenarios would be too high and they should therefore be considered to be *de minimis* risks.

[15]This suggests an explanation to the observation made by Allhoff (2009) that what is to count as a "catastrophe" in formulations of PP where this word is employed to signal the "great harm" part of the requirement of precaution is of great importance, although poorly understood.

connection is much less simple and straightforward than what is normally assumed in the *de minimis* debate.

2.3 Show

Dimension 2 can be said to denote a partly epistemic restriction – saying that an activity is subject to the restrictions set by the requirement of precaution only if there is some unspecified epistemic reason for believing that activity to lead to great harm. The epistemic classification is even more appropriate regarding dimension 3. This time, however, the epistemic restriction is active only *within* the limits set by dimension 2. For each activity meeting the criteria of dimension 2, the requirement of precaution says that something should be *shown* regarding that activity.

2.3.1 Proof-Standards

What this tells us is that the requirement prescribes us to produce some kind of (what in lack of a better word may be called) *proof* for the claim that an activity does not bring 'too serious risks' (i.e. given that this activity meets the condition set by dimension 2). But what is such a proof supposed to amount to, more precisely?

Although, in common usage, the word 'proof' is often equated with 'mathematical proof' or 'scientific proof', the request for proof made by the requirement can, in fact, be interpreted in very many ways. In the PP of the *Rio Declaration*, the need for scientific proof in order to justify precautionary measures is clearly discounted. Although this is hardly supported by available formulations, Hansson has even gone so far as to interpret this tendency as a categorical prohibition.[16] However, what this tells us is merely that the 'may'-part of dimension 2 may be interpreted very weakly – i.e. that scientific proof of great harm resulting from some activity is not needed in order for the requirement to be applicable. But this tells us nothing about what idea of valid proofs that should be built into the demands made by the requirement regarding such an activity. That is, it remains to be specified what kind of proof regarding the absence of 'too serious risks' that has to be produced in order for some activity *not* to be blocked by the requirement. Each such interpretation will specify what may be called the *proof-standards* set by the requirement of precaution.[17]

These proof-standards may, in turn, be more or less high. That is, they may vary in how much they demand for some statement to have been *shown*. However, the

[16]"[Waiting] for full scientific evidence is exactly what the Precautionary Principle tells us *not* to do" (Hansson 1999, p. 918).

[17]The term was introduced by Hansson (1999) and should not be confused with the notion of *evidence*. A proof-standard will, of course, have implications for what counts as a piece of evidence, but will also incorporate a basis for evaluating how various pieces of evidence should be assessed in combination, how the argumentative weight of a piece of evidence should be determined, rules for valid reasoning from evidence to conclusion, et cetera.

following seems to hold generally: all else being equal, the higher these proof-standards, the more difficult it will be to 'show' that some action will not bring too serious risks – and vice versa. In consequence, how high the proof-standards of the requirement of precaution are set will directly affect what degree of precaution it prescribes.

2.3.2 Decisional Paralysis

McKinney and colleagues have demonstrated that if a very weak interpretation of the 'may'-part of dimension 2 is combined with an interpretation of dimension 3 that sets very high proof-standards, the result may be a requirement of precaution that in many cases prohibits all alternatives open to the acting party in question (including the alternative of 'doing nothing').[18] This is due to the combination of the facts that if dimension 2 is made weak enough it will include all options and that, for none of these, it will be possible to 'show' that it will not 'bring too serious risks' (due to the difficulty of doing this resulting from the high proof-standards).

What this means is that certain interpretations of the requirement may systematically produce what will henceforth be called *decisional paralysis*. That is, the acting party addressed by the requirement will be recommended to abstain from all options open to him – which, of course, is an impossibility. Systematic production of decisional paralysis would, therefore, constitute a serious flaw of the requirement of precaution, since it would deprive it of any capacity to guide decisions and choices (and thus deprive PP of any chance of achieving direct political significance). Moreover, on a more theoretical level, if a norm recommends abstaining from every alternative in a situation, this seems to constitute a violation of the Kantian dictum that 'ought' implies 'can', i.e. the generally accepted claim that if an activity is to be wrong, irrational, impermissible, forbidden etc. it must be possible for the party that may undertake this activity to abstain from doing so. To be true, in the case of decisional paralysis, the acting party *can* abstain from *each single* prohibited activity *taken by itself*. However, it is impossible to comply with the conjunction of all such prohibitions issued by the requirement (since this conjunction tells the acting party to abstain from all options open to him).

McKinney and colleagues use this problem as a justification of the claim that PP is a hopeless idea that must be abandoned. However, rather than constituting that kind of knock-down argument, in my view, the possibility of decisional paralysis is at best a good reason for excluding this type of interpretations of the requirement of precaution. In effect, it may be suggested that a desideratum for interpreting the requirement is that, other things being equal, the higher proof-standards, the stronger also the interpretation of the 'may'-part of factor 2 – and vice versa.

[18] McKinney (1996), and McKinney and Hamer Hill (2000). Criticism along similar lines has later been presented in Harris and Holm (2002) and Sunstein (2005).

The just said also sheds some additional light on the argument from decision costs. For what has just been indirectly demonstrated is that, just as decisional paralysis is the upshot of a *combination* of certain interpretations in the 'may'- and 'show'-dimensions, this goes for the level of decision costs as well. In other words, the appeal to unacceptably high decision costs cannot in itself be an argument for interpreting any single one of these dimensions in some specific way. For any interpretation in the 'may'-dimension which, given some proof-standards, would give a certain level of decision costs, there is some weaker proof-standard such that, given the interpretation in the 'may'-dimension under consideration, the decision costs would be lower. For this reason, these costs can at best motivate certain *combinations* of such interpretations. This, in turn, is a result of the just demonstrated fact that these costs are very much dependent on the difficulty of meeting a certain interpretation of the requirement of precaution, and that this difficulty is influenced by both of the dimensions 2 and 3.

2.3.3 The Holistic Nature of Precaution

However, things are even more complicated. For if we now briefly consider also dimensions 4 and 5, it is quite easy to see that the ease or difficulty of meeting the requirement of precaution is also dependent on how it is specified in these respects. That is, what degree of precaution is actually prescribed by the requirement and whether or not this degree is so high that it results in unacceptable decision costs, or maybe even decisional paralysis, is determined by the specification of *all* dimensions 2 to 5 *taken together*. In this way, the degree of precaution prescribed by the requirement (and thus what decision costs and price of precaution it tells us to accept) must be seen as *holistically* determined by a sort of 'organic whole' consisting of specifications of the requirement in all of the dimensions 2–5.

For any specification of the proof-standards, the degree of difficulty of 'showing' that an activity does not bring 'too serious risks' may vary considerably with the choice of categorisation of such risks. If 'too serious risks' is interpreted in a way that makes it very difficult to show anything about such things with higher proof-standards, it may still be easier to show something with lower proof-standards. However, even if we adjust the interpretations in dimensions 2 and 3 according to the desideratum proposed above, the specifications in dimensions 4 and 5 may still lead to a situation where the requirement leads to decisional paralysis due to the fact that 'too serious risks' is taken to designate a type of facts about which it is very hard to gain any knowledge even with very low proof-standards. In all, therefore, the desideratum proposed above is not sufficient.

One possible response to this could be to attempt the formulation of a more complicated desideratum that excludes certain combinations of how the requirement is interpreted in all dimensions 2–5 (in much the same manner as the earlier suggested desideratum did regarding 2 and 3). However, not only would this require quite a lot of theoretical complication and, presumably, technical sophistication. Due to the very same holistic nature of precaution that creates the need supposed to be filled by such a construct, I am quite sceptic to the actual fruitfulness of such an undertaking. After all, the point of the present investigation is not to produce a list of unsound versions of the requirement in order to avoid these, but to find some version of the requirement that can be shown to be acceptable and justified. If it is required of such a version that it avoids decisional paralysis, the important thing is not exactly *how* this is achieved, but *that* it is achieved.

In consequence, while in one sense complicating the picture, the holistic nature of precaution actually makes the primary quest of this book somewhat easier to grasp. Keeping our awareness of the fact that decisional paralysis may result from many different variations in the various dimensions of precaution, we can rest content with the suggestion that any interpretation of the requirement of precaution that systematically leads to decisional paralysis is a flawed interpretation.

2.3.4 Conservatism and Arbitrariness

Besides manipulating the dimensions of precaution, decisional paralysis may also be avoided by applying a type of restriction to the requirement's range of applicability that was not considered in connection to dimension 2 above. If we let one of the alternatives in any situation of choice be immune to the restrictions set by the requirement of precaution, this can always function as a 'saviour's gate' out of decisional paralysis.

For example, McKinney and colleagues have considered the possibility that PP should not be applicable to the 'do nothing'-alternative (or, rather, the 'do not do things differently than before'-alternative) in any situation of choice.[19] This suggestion can easily be applied also to the requirement of precaution. Simply put, the option of preserving status quo is never among those activities that can be assessed from the requirement. If such a conservative approach is accepted, setting limits to what interpretations in dimensions 2–5 that may be accepted is unnecessary. Decisional paralysis may always be avoided due to the availability of the unconstrained 'do nothing'-option.

McKinney and colleagues reject this idea and, in my view, rightly so. While we might, in particular cases, have good reasons to adopt some kind of conservative approach, the conservatism expressed by the present idea is too *simplistic* since it gives status quo a universally privileged position *for no better reason than it being status quo* (i.e. for no reason at all). Moreover, since leaving things as they are may be just as dangerous and risky as any attempt to change status quo, such simplistic conservatism clearly runs contrary to the spirit of the underlying ideal of the desirability of precaution assumed to be expressed by the requirement.

---

[19] McKinney and Hamer Hill (2000). In my own experience, this 'favouring of status quo' is a rather common feature of many people's intuitive and spontaneous interpretations of PP.

However, the flaws of this approach go deeper than mere rejection of simplistic conservatism. For the problem with this kind of interpretation that has just been demonstrated is not merely that it gives status quo a privileged position, but that it does so without any hint of argument or reason. In particular, it lacks support from the underlying ideal of the desirability of precaution. In this sense, the simplistic conservative interpretation suffers from the flaw of being *arbitrary*.

Now, it seems to me that *any* interpretation of the requirement of precaution that, in a manner similar to simplistic conservatism, arbitrarily recommends *any* type of option should be rejected. This would exclude not only simplistic conservatism. Also its mirror image of, what might be called, simplistic worship of the novel (i.e., preferring new kind of activities and things just because they are new) suffers from the same flaw. And so do, it appears, all possible suggestions in between these two extremes that recommend some type of option for no better reason than the fact that it is of this type. The whole idea with PP and the requirement of precaution is that it is supposed to tell us something of normative importance – that acting against the requirement is somehow faulty. This, I conjecture, forces the requirement to suggest or (at least implicitly) imply *reasons* for its recommendations – i.e. arguments for why a recommendation is made that go beyond mere repetition of this recommendation.

I will return in Chapter 6 to the issue of whether or not some kind of conservative approach might be justified from the requirement of precaution, at least from time to time. For now, it suffices to note that what has just been formulated is a second desideratum of an acceptable interpretation of this requirement: *Besides avoiding decisional paralysis it should also avoid the issuing of arbitrary recommendations.*

2.4 Risk

Let us now take a closer look at the dimensions 4 and 5. As indicated by the brief sketch of the holistic nature of precaution, both these dimensions play a role in the determination of what degree of precaution is prescribed by the requirement. In effect, how the requirement is interpreted in both of these dimensions may influence whether or not it implies acceptable decision costs and thus prescribes a proper price of precaution. Moreover, in their extreme variations, they may both contribute to the requirement giving rise to decisional paralysis. However, more exactly *how* these dimensions might be varied as to produce such effects has not yet been made clear.

2.4.1 Likelihoods, Values or Combinations?

The basic concept of risk can be contrasted with our notion of actual harm or loss of value. The concept of risk employs the latter concept, but also involves the possibility of the harm actually never ensuing. When harm is effected, this harm actually occurs. However, when a risk of harm is effected, this is compatible with the fact that the risk is in fact never actualised – i.e., that no harm actually occurs. As will be seen later on (e.g. in Chapter 4), this basic structure of the concept of risk has important consequences for various arguments on the issue of how the imposition of risks may or may not be justified.

Besides this basic feature, however, the concept of risk may be interpreted in many different ways.[20] Here we encounter three traditional conceptions: (i) risk as the likelihood of a possible negative outcome, (ii) risk as this outcome itself, and (iii) risk as a combination of these two. Moreover, what is identified as a risk in these conceptions also corresponds to what is taken to determine the *magnitude* of a risk: (i) degree of likelihood of a possible negative outcome, (ii) degree of undesirability of such an outcome, and (iii) a combination of these.

The choice between these interpretations is important in the present context for the reason that the chosen interpretation has to be adequate in relation to the normative restriction set by the requirement of precaution not to impose *too serious* risks. The concept of risk must therefore be such that this restriction on risk impositions makes at least some sense from a normative point of view. Obviously, in order to determine whether or not that is the case, the idea of risks being too serious must be subjected to more in-depth analysis. In the following chapters, I will argue that the combinatory approach transpires as the most normatively relevant concept of risk from this point of view.

However, assuming that the concept of a risk can be adjusted to serve the need for normative adequacy still leaves important issues open. For, as indicated above, how the concept of risk employed in the requirement is understood in other respects may highly influence the degree of precaution prescribed.

2.4.2 Quantities, Qualities and Levels of Precision

One such aspect is what *level of precision* we require in the description of a risk. For example, should we interpret the terms 'likelihood' and 'undesirability' (or their counterparts in different formulations, such as 'probability' and 'value') in quantitative or qualitative terms? Perhaps, the most natural solution to that problem is to see these terms as denoting qualities that can be present in different quantities. But that only moves the problem to the issue of what level of precision we should assume for these quantities.

As observed by Hansson, if the likelihood or undesirability of a scenario is taken to denote only numerical probabilities or values, the requirement's range of applicability is restricted to those situations of choice where such numbers can be given (in a meaningful way).[21] The other side of this is that if the term 'risk' is taken to denote only likelihoods or values that can *only* be described in *non*-numerical terms,

[20] See, for example, Hansson (1989).

[21] Hansson (1997).

this excludes the applicability of the requirement of precaution to situations where numerical descriptions *are* possible.

What, then, transpires if we relate this unclarity to what degree of precaution is prescribed by the requirement? This, it seems to me, is initially a bit ambiguous. On the one hand, if this concept is narrowly interpreted according to any of the lines just described, the requirement's range of applicability will be restricted to certain situations of choice (and only some of the options in these situations). At first glance, this would seem to give a lower degree of precaution.

On the other hand, both of these interpretations present us with peculiar but different difficulties of justification. In one sense, showing the presence or absence of risks understood in numerical terms is much more difficult, since we have to have access to reasons that are exact enough to be able to justify very exact beliefs. In another sense, however, it is the idea of risks being present in *non*-numerical quantities that produces the greatest challenge from an epistemic point of view. For, if it is impossible to show the presence or absence of precise and well-defined quantities of risk, it becomes a bit hazy how one could ever justify a whole set of central types of beliefs regarding risks: For example, beliefs regarding the presence of *a* risk rather than *no* risk (and vice versa), *this* risk being *greater or lesser* than *that* risk, or a risk being greater or lesser than *this or that (non-numerical) quantity* of risk.

In all, therefore, appeal to the conditions of avoiding decisional paralysis and prescribing an acceptable degree of precaution cannot really decide this matter at the present stage of inquiry. Whether risk magnitudes should be understood as numerical or non-numerical quantities or both therefore remains an open question from this point of view. Moreover, as will be seen in later chapters, many of the reasons that may be mobilised in support of the requirement of precaution do not distinguish between situations of choice or activities along any of the lines just sketched. That is, many of these reasons seem to imply that precaution may be desirable quite independently of whether or not the risks actualised can be described in numerical terms or not. In my view, therefore, it would be unwise to restrict the requirement's range of applicability in any of these ways unless its justificational basis provides good reason for such a restriction.

One important consequence of all this is that identifying the requirement of precaution (or PP for that matter) with the so-called *maximin* rule[22] is not very fitting as point of departure for an analysis.[23] The reason is that this rule is commonly understood as a complement to standard decision rules[24] in situations where these latter rules are useless due to lack of numerical likelihoods and values.[25] But if, as has just been argued, many of the reasons that may be employed in support of the

---

[22]This rule tells us to focus on the avoidance of very bad outcomes and therefore instructs us to choose that option the worst possible outcome of which is at least not worse than the worst possible outcome of every other option open to us. See further the discussion of this rule in Chapter 3 below.

[23]See, for instance, Hansson (1997) for an illustration of such an assumption being made.

[24]Typically the principle of maximising expected utility, explained in Chapter 3.

[25]See, e.g., Rawls (1971, pp. 152ff.), and Resnik (1987, chapter 2).

requirement do not limit its range of applicability in this way, it would seem unwise to simply assume these reasons to be invalid.

As a starting point, therefore, I will understand the concept of risk employed in the requirement of precaution widely. Both numerical and non-numerical descriptions of likelihoods and degrees of undesirability of outcomes are included.

2.4.3 Objective or Subjective?

A third unclarity in the concept of risk concerns its ontological implications – what *kind* of entities risks are supposed to be. Are risks objects in the world (or properties of such objects), beyond our thoughts and other attitudes? Or are risks nothing more than a certain subset of precisely these thoughts and attitudes? Or are risks perhaps a bit of both?

This well-known 'objective-subjective' debate seems to bear some relevance to the issue of how to interpret the requirement of precaution, since it makes a difference as to *what* one is supposed to show the absence of when showing the absence of too serious risks. It is not hard to imagine that it may be more or less difficult to undertake this depending on whether risks are subjective psychological properties or properties of the world independent of people's minds (or some kind of mix). In effect, the choice of interpretation regarding this aspect seems to matter for what degree of precaution will be prescribed by the requirement.

It should be observed that most practical dealings with risks seem to employ a combination of the subjective and objective conceptions. In the social management of risk, for example, it is commonplace to rely on risk estimates made by people (the subjective conception), but at the same time require some evidence that supports these estimates[26] (the objective conception). In Chapter 5, I will argue that there are moral reasons for employing such a mixed conception in relation to the requirement, since this facilitates plausible moral claims with regard to the quality of the evidence underlying risk-assessments.

2.5 Too Serious

This takes us, finally, to dimension no. 5 – the idea of risks being 'too serious'. Apparently, the central unclarity here regards where, on the scale of risk-magnitude, we should locate the 'too serious'-threshold (if, indeed, there is a threshold rather than a vague 'grey area'). As was indicated above, all else being equal, the lower this threshold, the greater the degree of precaution prescribed by the requirement – and vice versa. That is, the placing of this limit on the scale measuring risk-magnitude

[26] Or, at least, that the estimates are made by people that are seen as especially reliable with regard to the matter at hand (i.e. experts).

will directly influence what price of precaution the requirement prescribes us to pay and even – in its extreme variants – threatens to produce decisional paralysis.

However, things are even more complicated. For it is by no means a given that the *seriousness* of risks is a simple function of their *magnitudes*. First, as we saw above, this partly depends on what risks are taken to be and what, by virtue of this, *determines* their magnitude. For example, if the magnitude of a risk equals the mere *likelihood* of a possible negative outcome, many would presumably hold that a risk might be quite serious although its magnitude is rather slight. This since the outcome that would effect an actualisation of the risk may still be very undesirable. Similarly, if the concept of risk is understood purely in subjective terms, it can be questioned why the magnitude of risks (which in this case equals the private beliefs and attitudes of individual people, regardless of if these are the results of pure guesswork, mental instability, wishful thinking or rigorous critical scientific inquiry) should carry much normative weight at all.

Secondly, regardless of which of the above-mentioned conceptions of risk are used, it may be claimed that the restriction set by the requirement of precaution should also be guided by *other* factors than mere risk magnitude. Suggestions made in this direction point to as possible candidates either some qualitative property of the outcome aspect of the risk (such as if the outcome is committing an act of murder, or the extinction of humankind[27]), or some relational property holding between the risk itself and the people on whom it is imposed (such as the risk being unwished for, or being unfairly distributed[28]).

It may be argued that at least some of these additional aspects have little to do with the intuitive idea of the desirability of *precaution*, but rather with *other* normative ideas. Again, however, the wise thing to do would seem to leave it open whether or not such a tightening of the concept of precaution should be adopted. After all, if certain aspects of what is being risked in some activity (for example, that this activity brings a very small likelihood of the extinction of humankind) is so important that this activity should therefore be avoided altogether, this would seem to have significant effects on the degree of precaution we are obliged to exercise. Perhaps this increase cannot easily be traced back to any basic idea of the desirability of precaution *as such*. Nevertheless, however, it would be a product of a more general family of ideals regarding what risk impositions are acceptable – a family of which the requirement of precaution is obviously a member.

In all, therefore, it seems that the unclarities of this final dimension open up a promising perspective for the forthcoming inquiry. Although the central issue will

[27] The former has been repeatedly suggested by Catholic officials and ethicists in debates on the morality of abortion and embryo research, see, e.g., Ford (1990), and Mahoney (1984, p. 68), but also by proponents of animal rights, see Bradshaw (1998). The latter is the suggestion made by German theologian Hans Jonas' so-called *imperative of responsibility*, see Jonas (1979).

[28] The former idea seems to have originated in Nozick (1974, pp. 73–76), and has then been pursued in different ways by commentators such as McCarthy (1997), Shrader-Frechette (1991), Teuber (1990), and Thomson (1986). The latter idea has been pursued by, among others, Perhac (1999), and Schuyt (1998).

remain the degree and proper price of precaution, we have caught a glimpse of how this issue may be connected to other issues regarding what is of value and what we should do. For example, if the extinction of humankind would be an extremely bad thing – or a thing that we are strongly obliged to avoid if we can – this may be seen as a reason for accepting a very high price of precautionary measures designed to prevent such an outcome.

2.6 Summing Up

Several important preliminary conclusions have been reached in this chapter.

First, two desiderata for an acceptable interpretation of the requirement of precaution have been formulated. Such an interpretation must not systematically produce decisional paralysis, and it must avoid making arbitrary recommendations.

Secondly, we have seen that there is virtually no end to the number of ways in which the original formulation of the requirement of precaution may be varied in its various dimensions of unclarity. The degree and price of precaution prescribed by the requirement is in this way connected to a whole web of issues in decision and risk theory, epistemology and the philosophy of science, as well as moral and political philosophy. While in one sense severely complicating the quest pursued in this book, I have argued that the main effect of this complexity is, in fact, a greater clarity and simplicity as to what is really important in the clarification of the requirement.

This argument connects to the third conclusion of this chapter regarding the holistic nature of precaution. What this means is that the complexity just outlined in many ways makes it unnecessary to consider all those details that would figure in any precise version of the requirement. All these details can be seen as a large set of conceptual levers which in different combinations may produce more or less satisfying results in terms of the degree and price of precaution prescribed by the requirement (as well as the two 'formal' conditions mentioned above). Moreover, to what extent the setting of one of these levers will actually influence these factors is wholly dependent on how all the other levers are set. In consequence, for each degree and price of precaution, these can be prescribed by the requirement of precaution as a result of a multitude of *different* combinations of settings of these levers. The important thing, therefore, is not exactly how all these levers are set – i.e. how the requirement is interpreted and formulated in detail – but that they are set in *some* way that yields an acceptable result.

This takes us to the fourth and last conclusion regarding the actual fruitfulness of the fact that the requirement is in so many ways connected to a host of issues in various fields. For if, in the following inquiry, we do not need to bother with the fine details of producing precise formulations of the requirement, we can concentrate wholeheartedly on the primary issue of what makes for an acceptable degree and a proper price of precaution. And, in that investigation, we are, in fact, strongly aided by the connections between the unclarities of the requirement and more well-established areas of inquiry, since we can turn to these in our search for arguments regarding what makes for an acceptable degree and a proper price of precaution. Therefore, it is to these areas that we will now turn in order to keep pursuing the quest for a normative basis of precaution.

References

Allhoff F. "Risk, Precaution, and Emerging Technologies." *Studies in Ethics, Law, and Technology* 3 (2) (2009). doi: 10.2202./1941-6008.1078.

Andersson P. "Humanity and Nature: Towards a Consistent Holistic Environmental Ethics." PhD diss., Göteborg: Acta Universitatis Gothoburgensis, 2007.

Arrhenius G. "An Impossibility Theorem for Welfarist Axiologies." *Economics and Philosophy* 16 (2) (2000): 247–66.

Bradshaw R.H. "Consciousness in Non-human Animals: Adopting the Precautionary Principle." *Journal of Consciousness Studies* 5 (1) (1998): 108–14.

Brülde B. *The Human Good*. Göteborg: Acta Universitatis Gothoburgensis, 1998.

Ford N.M. "Ethics, Science and Embryos." *The Tablet*, February 3 (1990): 141–2.

Hansson S.O. "Adjusting Scientific Practice to the Precautionary Principle." *Human and Ecological Risk Assessment* 5 (5) (1999): 909–21.

Hansson S.O. "Dimensions of Risk." *Risk Analysis* 9 (1989): 107–12.

Hansson S.O. "The Limits of Precaution." *Foundations of Science* 2 (1997): 293–306.

Harris J. and S. Holm. "Extending Human Lifespan and the Precautionary Paradox." *Journal of Medicine and Philosophy* 27 (3) (2002): 355–68.

Jonas H. *Das Prinzip Verantwortung. Versuch einer Ethik für die technologische Zivilisation.* Frankfurt am Main: Insel Verlag, 1979.

Mahoney J. *Bioethics and Belief*. London: Sheed & Ward, 1984.

McCarthy D. "Rights, Explanation, and Risk." *Ethics* 107 (1997): 205–25.

McKinney W.J. "Prediction and Rolston's Environmental Ethics: Lessons from the Philosophy of Science." *Science and Engineering Ethics* 2 (4) (1996): 429–40.

McKinney W.J. and H. Hammer Hill. "Of Sustainability and Precaution: The Logical, Epistemological, and Moral Problems of the Precautionary Principle and Their Implications for Sustainable Development." *Ethics and the Environment* 5 (1) (2000): 77–87.

Mumpower J. "An Analysis of the de Minimis Strategy for Risk Management." *Risk Analysis* 6 (1986): 437–46.

Nozick R. *Anarchy, State and Utopia*. New York: Basic Books, 1974.

Parfit D. *Reasons and Persons*, 2nd printing. Oxford: Clarendon Press, 1984.

Perhac Jr. R.M. "Environmental Justice: The Issue of Disproportionality." *Environmental Ethics* 21 (1) (1999): 81–92.

Peterson M. "What Is a *de Minimis* Risk?" *Risk Management: An International Journal* 4 (2) (2002): 47–55.

Rawls J. *A Theory of Justice*. Oxford: Oxford University Press, 1971.

Resnik M.D. *Choices: An Introduction to Decision Theory*. Minneapolis & London: University of Minnesota Press, 1987.

Sandin P. "Naturalness and *De Minimis* Risk." *Environmental Ethics* 27 (2) (2005): 191–200.

Sandin P., M. Peterson, S.O. Hansson, C. Rudén, and A. Juthe. "Five Charges Against the Precautionary Principle." *Journal of Risk Research* 5 (4) (2001): 287–99.

Schuyt K. "The Sharing of Risks and the Risks of Sharing: Solidarity and Social Justice in the Welfare State." *Ethical Theory and Moral Practice* 1 (3) (1998): 297–311.

Shrader-Frechette K. *Risk and Rationality. Philosophical Foundations for Populist Reforms.* Berkeley: University of California Press, 1991.

Shrader-Frechette K. "Technological Risk and Small Probabilities." *Journal of Business Ethics* 4 (1985): 431–46.

Stenmark M. *Environmental Ethics and Policy-Making*. Aldershot: Ashgate, 2002.
Sunstein C.R. *Laws of Fear: Beyond the Precautionary Principle*. Cambridge: Cambridge University Press, 2005.
Teuber A. "Justifying Risk." *Dædalus*: *Proceedings of the American Academy of Arts and Sciences* 119 (4) (1990): 235–54.
Thomson J.J. *Rights, Restitution and Risk*. Cambridge, MA: Harvard University Press, 1986.
Weinberg A.M. "Science and Its Limits: The Regulator's Dilemma." *Issues in Science and Technology* 2 (1985): 59–72.
Whipple C. ed. *De Minimis Risk*. New York: Plenum Press, 1987.

Chapter 3
Precaution and Rationality

The analysis undertaken in the foregoing chapter made clear that one area of particular interest in relation to the various dimensions of precaution is the field of decision theory. In this field, concepts of central importance for the requirement of precaution such as risk, probability and value – as well as ideas regarding the determination of magnitudes of these parameters – are the subject of sophisticated theorising. More importantly, however, in this field, these concepts are used for explicit *normative* inquiries regarding the *rationality* of actions and decisions. And, as we will see, ideas regarding this have been used as arguments in favour of precautionary suggestions. Therefore, it is of interest to investigate to what extent they may be taken as a point of departure in our quest for a normative basis of precaution.

In order to be able to explain how this might be done, we must first sketch in some more detail than what has been done previously the standard view of rational acting that forms the basis of most of the theorising within decision theory. Such an overview may be a bit tedious for those who are already familiar with the field. Still, this short presentation will focus particularly on aspects of the normative discussion of rational action that I will refer to and expand on in later parts of this book, or have simply found to have the most relevance for the issue of precaution in general.

3.1 Rational Action – the Standard View

The basic idea within decision theory regarding what makes an action rational or irrational can be expressed something like this: If you know for certain what is worth striving for, what options you face in a particular situation and which of these options that would best realise that which is worth striving for, the rational thing to do in this situation is to choose this option. All other decisions would, in these circumstances, be irrational. This is what decision theorists refer to as the rationality of decisions under conditions of *certainty*.

C. Munthe, *The Price of Precaution and the Ethics of Risk*, The International Library of Ethics, Law and Technology 6, DOI 10.1007/978-94-007-1330-7_3,

3.1.1 Efficiency, Value Neutrality and Calculated Risk Taking

Already here, we can see some of the central features of the standard view. It is an idea of rationality as *efficiency* or *productivity* – only the option that *maximises* whatever is of value deserves to be called rational. At the same time, it is in itself quite neutral as to what really *is* worth striving for – the core of rationality in the standard view has nothing to do with that, although it must assume that *some* such notion of value can be made to make sense. Following British eighteenth century philosopher David Hume's classic account of practical reason,[1] the more worked out theoretical accounts of the standard view expresses this latter feature by identifying rationality with efficient preference satisfaction or desire fulfilment – whatever you wish for or values the most it is rational to achieve.

However, even if we know what option would be rational had we only known for certain what option would maximise whatever is of value, in reality we are never choosing under such conditions. We may have some ideas as to which of our options would best achieve our goals. However, these ideas are not to be trusted. First, we may have false beliefs that, if we were to trust them, would lead us to choose inefficient options. Secondly, we might have gaps in our repertoire of beliefs, so that certain aspects of great importance for the efficiency of different options are completely unknown to us.

In the first case, we are in what decision theorists refer to as the condition of *risk*. The fact that we realise that our beliefs may be false still permits us to estimate their likelihoods or probabilities. And, according to the standard view, these likelihoods may be used for deciding what to do in a way that may still be called rational. The line of reasoning leading up to this expansion of the standard view may be summarised in the following way.

First, the rational way of deciding on the basis of likelihoods or probabilities cannot be just to choose the option that is *most likely to maximise value*. For example, such a notion would tell us that if presented with the offer of playing Russian roulette in order to win €1, the rational thing to do would be to accept this offer (since the most likely outcome of this is that we receive some additional money). Obviously, such a formula would miss out on the highly relevant aspect that what might produce most value might also instead produce most *dis*value. And since the basic idea of efficiency expressed by the standard view also takes into account the importance of avoiding losses, the probability of success has to be balanced against such risks of losing it all.

However, and secondly, it would be just as irrational to fall for the simplistic temptation of making this latter aspect one's only concern – i.e. just choosing the option where the worst possible loss is least likely. For one thing, this option might still be the one where all *other* possible losses are *most* likely. More importantly, however, this line of reasoning minimises the chance of receiving any type of gain – and this regardless of the ratio between possible gains and losses. For example,

---

[1] Hume (1978, book II, part 3, section 3; book III, part 1, section 1).

presented with the offer of buying a ticket in a lottery for €1 with a 99% probability of winning €1,000,000, we should decline in order to escape the 1% risk of ending up €1 short. Obviously, this line of reasoning misses out on the central feature of the standard view that, besides avoiding losses, it is also highly rational to pursue the prospect of receiving gains.

Together, these two flawed ideas about how to rationally decide on the basis of likelihoods point to the need for considering the whole spectrum of both *all the likelihoods* of possible outcomes and *all the values and disvalues* of these outcomes. Therefore, estimates of likelihoods are rationally utilised in decision making only if these are balanced against each other and against the values and disvalues of the different possible outcomes. In that case, we can rationally decline the Russian roulette offer by pointing to the apparent fact that receiving a five to six chance of an additional €1 cannot outweigh running a one to six risk of losing one's life.[2] However, taking risks in order to receive gains may be a rational thing to do if the balance of values and probabilities are more in one's favour. For example, taking a 1% risk of losing €1 in order to receive a 99% chance of winning €1,000,000 can be judged as rational due to the apparent fact that the greatest possible loss in this case is not much of a loss, and even less so if considered in the light of its probability and compared to what will be gained in 99 times out of 100 if the offer is accepted.[3] This is the basic idea of, what I will often refer to as, *calculated risk taking* – expressed in a very exact way by the standard rule within decision theory for rational action under conditions of risk: the *principle of maximising expected utility*.

This principle assumes that, under conditions of risk, we can specify numerical values for both the probability and the value of the various possible outcomes of available options. For each such option, the probability, p, and the value, v, of each possible outcome can then be multiplied into a product, PR. For each option, the PRs of all its possible outcomes can then be added into a sum, EU, which is the expected utility of this option. The principle of maximising expected utility then prescribes that the rational decision is to choose the option with the greatest EU.

Following this formula, we can now analyse the situation where we decide whether or not to accept the offer of Russian roulette on the assumptions that winning €1 is of some but very slight value, represented by the number 1, while losing one's life is of great disvalue, represented by the negative number –100:

[2]This, of course, assumes a certain evaluation of the outcomes. If a person values very highly to receive some extra money but does not have much against losing his life, it may still be rational to accept the offer.

[3]Again, this line of reasoning assumes a certain evaluation of the outcomes. If the €1 I could use to purchase a ticket in the lottery is the only money I have and I will starve to death if I do not soon get some food (which could be purchased for the €1), opting out on the offer of entering the lottery may seem as the rational thing to do in spite of the very high probability of winning.

Option 1: Accept ➔ 83% chance of winning ➔
PR = p: 0.83 × v: 1 = 0.83
➔ 17% chance of losing ➔
PR = p: 0.17 × v: –100 = –17
EU = 0.83 + (–17) = –16.17

Option 2: Decline➔ 0% chance of winning ➔
PR = p: 0 × v: 1 = 0
➔ 0% chance of losing ➔
PR = p: 0 × v: –100 = 0
EU = 0 + 0 = 0

Since zero is a greater number than –16.17, option 2 has the greater expected utility. Hence, the principle of maximising expected utility prescribes that the rational thing to do is to decline the offer of playing Russian roulette.

3.1.2 Enlightment Critique and the Charge of Instrumental Rationality

In the first chapter, I remarked that the principle of maximising expected utility constitutes the theoretical basis of contemporary standard models of risk analysis and rational policy making. Furthermore, these models seem to express the very core of the contemporary version of the enlightment ideal – i.e., the ideal of a society designed through rational planning powered by scientific knowledge. Based on the more detailed account of the standard view just undertaken, its apparent connection to the enlightment ideal can now be explained in more detail. This is of some interest, since it is highly unclear whether PP – and, in effect, also the requirement of precaution – prescribes a deviation from this ideal or not. And to the extent that it does, it is still unclear whether or not this would force us to abandon contemporary standard models of risk-analysis and rational policy making.

The message of efficiency or productivity of the standard view can be taken immediately to heart by the enlightment ideal, since that is as much a part of its own core. The idea of calculated risk taking brings scientific thinking and facts right into the very core of the standard view of rationality[4] – quite in tune with the enlightment ideal. The aspect of value neutrality makes for a formula adaptable to whatever political goal that may be considered – a highly important characteristic for a political ideal where democracy, with its shifting opinions and majorities on what goals society should pursue, is seen as the final solution to the question of who should be the evaluative authority of society. At the same time, in this last aspect,

[4]Both through the application of mathematical, statistical and scientific models of probability, measurement and the aggregation of values, and through the need for scientifically organised information in order to be able to specify what options we face in various situation, what outcomes these options might produce and the probabilities of these outcomes.

hecklers of these ideas have spotted a powerful piece of rhetoric – the accusation of the standard view to present (and the enlightment ideal to be based on) a theory of mere *instrumental* rationality.[5]

In relation to my discussion, this kind of criticism has some but still limited concern. First, it would seem that supporters of the standard view could easily accept the idea of a need for some kind of supplement to this view that addresses the acceptability of goals and values. That is, they could say that, just because an action is *rational*, it need not be recommendable or permissible all things considered (since what it maximises may not be worth pursuing). Second, exactly this kind of addition is obviously in place when it comes to our present inquiry. As mentioned in Chapter 2, one of the central dimensions of precaution concerns what underlying theory of value should inform the requirement of precaution. What this means is that, although an activity would have been permitted by the requirement had we considered one set of values (for example, effects on human well-being), another set of values (for example, one including also the well-being of other sentient creatures) may instead have made the requirement prohibit this activity. However, seen from the point of view of the standard view of rational action, such differences have nothing to do with the *rationality* of this activity. That is, a theory of 'instrumental rationality' of this kind does not really bring forward any claim regarding what activities should be pursued all things considered, but merely concerns itself with *one* factor that may be of relevance for this (i.e. the rationality of the various options). In effect, then, the alleged flaw of instrumentality would not seem to be a reason for any radical shift in our standard view of practical rationality. Rather, it is the criticism holding out this alleged flaw that presupposes a flawed interpretation of the standard view – at least in relation to its more contemporary and advanced forms.

This view regarding the normative ambitions of the standard view is, as will soon be demonstrated, of some importance for the quest for a normative basis of precaution. In the sense of having inspired the above described moderation of these ambitions, the criticism based on the charge of instrumentality therefore has some relevance for our discussion. However, it does not seem to be the case that this accommodation to typical enlightment critique really forces the requirement *entirely* out of the context of the classic enlightment ideal and its underlying conception of rationality. Rather, what we will see is that reasons of rationality have to be amended by *other* normative reasons regarding what activities are defensible

[5]This term is taken from German philosopher Jürgen Habermas, who in the spirit of Immanuel Kant has formulated objections along this line to the enlightment ideal and its underlying theory of rationality (or, as Habermas would have it, 'the philosophical discourse of modernity'). Although undertaken from different historical standpoints (Kant from within the very birth of the enlightment, Habermas from its matured version of the modernity of the late twentieth century) and although inferring quite different conclusions from this criticism, they both point to the arbitrariness of what goals are to be pursued in the enlightment version of rationality as its most fundamental flaw. See Kant (1998) and Habermas (1991), respectively. For further and more recent criticism regarding the instrumentalist element of the standard view of rational action, see Parfit (1984, chapters 6–9), and Smith (1994).

all things considered. However, although being insufficient for justifying any fully fledged version of the requirement of precaution, instrumental rationality in the form of the standard view, still has important elements and ideas to offer our pursuit of a normative basis of precaution.

3.2 Rational Precaution

Besides the charge of instrumentality, there are a lot of different objections to the principle of maximising expected utility in other respects. Most of these objections can still be localised *within* the basic standard view of rational action and the general approach of calculated risk taking. This, for example, is the case with the criticism underlying the idea of *de minimis* risk mentioned in the preceding chapter. Belonging to a family of objections according to which strict application of the principle of maximising expected utility as a practical method in every situation of choice will in itself not be an option that maximises expected utility,[6] this criticism is of interest mainly through its insistence on greater complication and sophistication of the standard view and the calculated risk taking approach. However, in itself, this family of objections suggests no alternative basic conception of rationality. The same can be said about those numerous examples where, apparently, people who seem to be perfectly normal are not disposed to choose or deliberate according to the principle of maximising expected utility.[7] While showing that, probably, this principle is in many cases not a very reliable predictor of actual human thinking and behaviour, they do not really challenge its basic conception of rational action. Rather, they seem to confirm what most people believe anyway: that, in many cases, we do not think or act fully rational.

3.2.1 Ignorance, Precaution and the Maximin Rule

In order to find more substantial challenges to the standard view, we have to move beyond the framework assumed so far and point to the fact that most actual situations of choice do not seem to involve decisions under conditions of risk. For, in reality, we are often afflicted not only by lack of certainty in our beliefs, but also by sheer *ignorance* or even *lack of opinion*. That is, we are unable to specify probabilities representing precise likelihoods of various possible outcomes. Perhaps, to some extent, we are able to give very rough qualitative estimates of likelihood (for

[6] The so-called St. Petersburg paradox is one classic example of this kind of objection. See Resnik (1987, pp. 107–108), for a basic presentation. A rather different argument in favour of the claim that the principle of maximising expected utility may give us reason not to act according to this principle in particular cases is the main theme of Gauthier (1986).

[7] The list of examples regarding this could be made very long. Classics in the decision theoretical literature are Allai's and Ellsberg's (so-called) paradoxes (see Resnik 1987, pp. 103–106, for a basic presentation).

example, like in the statement that some event is not wholly unlikely to occur). However, and more importantly, in many cases we will lack *all* knowledge regarding likelihoods. Either (perhaps mostly?) because we are ignorant about some of our options and the possibility of various outcomes, or because we are unable to say any more about the possible outcomes we know about than merely stating this 'possibility'. How would a rational maximiser of whatever is of value act under such circumstances if he is to honour the spirit of the standard view?[8]

One answer to this question is that rational action in circumstances like these would presuppose the transformation of the situation into a decision under conditions of risk. From this perspective, two roads seem to be open: either to acquire further information so that it becomes possible to calculate the expected utility of all options, or to find a true description of the original situation that permits such a calculation in spite of the apparent uncertainty or ignorance. I will comment on the first of these roads immediately and return to the second one in the following subsection.

From the perspective of the standard view, the acquisition of further information would be considered rational only if such an action would maximise expected utility. However, except for a few special cases,[9] to answer the question whether or not this condition would be met would require knowledge of what option would maximise expected utility in light of this additional information (so that we can know whether or not the same option would be chosen anyway – i.e., if the additional information is not acquired[10]). Unfortunately, however, given the circumstances of ignorance, to have such knowledge is impossible. In consequence, therefore, the standard view cannot say very much regarding the rationality or irrationality of collecting additional information in order to decrease ignorance.

---

[8]This aspect of risky decisions connects to what Peter Gärdenfors and Nils-Erik Sahlin have called *epistemic risk* – i.e., the probability that our assessment of risk is faulty in some respect in combination with the possible consequences of a decision in light of this. If, in most situations where we decide under conditions of risk, we are afflicted also by epistemic risks, this seems to mean that the real-world versions of the idealised idea of decision under risk are in fact closer to decision under uncertainty or ignorance. See, e.g., Gärdenfors and Sahlin (1982).

[9]Namely, in situations where it is possible to, first, represent our ignorance by very crude but nevertheless numerical probability intervals and, second, show that no more precise probability estimate made with the help of additional information would make a difference as to which option would maximise expected utility (or satisfy some other decision rule in need of exact probability estimates). See Malmnäs (1994, 1999). For a not too technical explanation of the basic features of probability theory that make such reasoning possible, see Resnik (1987, pp. 57–61).

[10]In fact, the situation is even more complicated. Since the very process of acquiring additional information will always require that resources and time are spent on this rather than some alternative activity, such acquisition will always increase the decision costs. Therefore, even if the option that would be chosen without this further information would have a lower expected utility than the one that would appear rational in light of the additional information *if these options are taken by themselves*, to acquire this additional information may still be irrational according to the standard view due to the fact that the expected utility of the latter option will have to be weighted down with the additional decision costs. That is, the additional expected gains made possible by the additional information have to be substantial enough to balance out the additional decision costs.

If, then, there is no rationally required strategy that could take us out of the condition of ignorance, it would seem that rational action would have to be undertaken within this framework. In such situations, many have argued, a rational value maximiser would start to think mainly about safety and concentrate on taking precautions in order to avoid suffering very bad losses. Much as a professional financial investor in times of recession will concentrate on avoiding complete ruin, the rational agent under conditions of ignorance will shift his attention away from the possibility of additional gains in order to concentrate on avoiding unforeseeable catastrophic outcomes.

Within the framework of decision theory, such a strategy may be represented in slightly different ways. However, I will here follow Sven-Ove Hansson's claim that the most plausible of these would be the suggestion that the criterion of rational action under conditions of ignorance is to decide according to the *maximin rule*.[11] This rule tells us that the rational decision is to choose that option, the *worst* possible outcome of which is not worse than any other option's worst possible outcome. That is, we need not bother with either the probabilities of various possible outcomes or exactly how desirable or undesirable these outcomes would be. All we need to do is to identify the worst outcome that may result from each option and then choose that option where the worst outcome would still be at least as good as the worst outcome of any of the other options.[12] For this reason, the maximin rule may seem to conform both to the intuition of the need for precaution under conditions of ignorance and to the need for a criterion of rational action in line with the standard view that can be applied under such conditions.

In spite of this, ignorance in itself does in fact not seem to constitute a *sufficient* reason for adopting a precautionary strategy. In fact, if we are to decide in the spirit of the standard view of rationality, ignorance might as well inspire us to choose according to the very *opposite* of the maximin rule and just go for the option with the *best* possible outcome (the so-called *maximax* rule). As observed by John Rawls in his famous analysis of the conditions and principles of justice, in order to provide a reason in line with the standard view for the application of the maximin rule, we also have to add the condition that substantial values are at stake in the choice. That is, besides being under conditions of ignorance, the choosing party is at risk of suffering significant harm or loss due to the choice he makes. This is the situation of the parties in Rawls' 'original position' (where, in Rawls' theory, a hypothetical choice of the basic constitutional principles of society is to be made): they know that they may turn out to be the losers in the society the basic structure of which they are designing, but they do not know *what* choice will have that effect. In such a predicament, Rawls argues, it is wise and rational in the spirit of the standard view to 'play it safe' and concentrate on avoiding the most catastrophic outcomes,

[11] Hansson (1997).

[12] Obviously, if several options have equally bad worst outcomes, the same reasoning can be repeated regarding the second worst outcomes, and so on.

even if that means giving up chances of extreme prosperity.[13] In simple terms, what Rawls suggests is that, in the face of ignorance where one runs the risk of significant loss, it may be rational to buy an insurance against the possibility of this loss taking extreme proportions (should it occur) for money that might instead have been used for highly enjoyable and certified pleasures.

3.2.2 Limitations of Plausibility, Applicability and Status

Strong as the case in favour of the maximin rule as a principle of rational action under conditions of ignorance may appear, there are also substantial reasons against this suggestion. As mentioned above, even if we cannot rationally justify any strategy out of ignorance, there may still be possible to find a way of representing the situation in a way that makes the calculation of probabilities and expected utilities possible. And, if so, the standard view clearly prescribes that the rational decision is to choose the option that maximises expected utility.

John Harsanyi has described such a strategy in connection to an example very close to the above described choice used by Rawls as a reason in favour of the maximin rule.[14] Again, the choosing party is to choose in what society to live in light of general knowledge of the conditions of this society but in lack of the crucial information of what position in this society he himself will in fact occupy. What Harsanyi suggests is that, although fatally incomplete, the information available to the choosing party may still be used as a basis for the calculation of probabilities. Just as in a lottery with 100 tickets and 1 winner, we have reason to infer that any randomly picked ticket has a 1/100 probability of being the winner, the choosing party can make a similar calculation of his probability of being any randomly chosen individual in the societies he ponders. And, given that he can value clearly the various positions occupied by these different individuals, he can use these 'lottery probabilities' in a further calculation of the expected utilities of the societies from which he is to choose. In effect, ignorance – even in combination with the risk of extreme harm or loss – provides no reason in the spirit of the standard view for applying the maximin rule or, for that matter, any other strategy that may deserve to be called precautionary.

Rawls, of course, does not accept this reasoning. He claims the probabilities in Harsanyi's solution to be purely statistical constructs with no substantial basis of empirical information. Instead, they result primarily from the so-called *principle of insufficient reason*, according to which we in lack of knowledge may infer probabilities in the way we do regarding the outcomes of lotteries and similar phenomena.[15] Not questioning the mathematical soundness of this principle, Rawls still insists

[13]Rawls (1971, pp. 152 ff.).

[14]Harsanyi (1976).

[15]More precisely, this principle states that if we lack reason to believe that any one of a number of outcomes is more probable, we may assume these outcomes to have equal probability. That is, if

that probabilities derived in this way cannot provide the basis for a calculation of expected utility potent of providing reasons for one option being *more rational* than others.[16] However, as much as one may sense that there is something to this claim, it is also hard to deny the power of Harsanyi's line of reasoning.[17] It seems, rather, that the disagreement is about subtleties regarding what really constitutes the spirit of the standard view of rationality – an issue where any certainty regarding what position is the most plausible one seems unreachable. In effect, the reason that may be provided for precaution (in the form of the maximin rule) from the point of view of the standard view of rationality can, in my view, at best be seen as rather weak and unstable.

In addition to this, the argument in favour of the maximin rule as a criterion of rational action described above severely limits its range of applicability in a way that seems quite problematic to take for granted in the present context. This weakness of the suggestion that the requirement of precaution (or PP itself for that matter) should be indentified with the maximin rule was briefly mentioned in the foregoing chapter and we can now see it more clearly. Within the context of the standard view of rational action, the best we can achieve with this strategy is a plea for precaution in situations when we face decisions under conditions of ignorance and/or uncertainty (and where there is no rationally required way out of these conditions). However, as has been demonstrated earlier and will be seen even more clearly in later chapters, neither the context of PP nor those normative perspectives that may justify some more precise version of the requirement imply that the basic ideal of the desirability of precaution must be assumed to be limited in this way.[18] Hence, the standard view begs substantial normative questions regarding the range of applicability of the requirement by assuming the idea that decisions under conditions of risk, uncertainty and ignorance have to be assessed according to different normative standards. It may very well be, when all normative reasons have been considered, that the requirement of precaution indeed prescribes us to act contrary to what would be considered rational under conditions of risk according to the standard view.

Admittedly, it would be possible to harmonise the standard view with such a requirement of precaution, if fulfilment of the prescriptions of the latter are assumed to be a (sufficiently large) value that must enter the calculation of expected utility.[19] However, the standard view itself implies no support for *requiring* such a thing. On

---

we consider two outcomes their respective probability is 0.5, if they are three it is 1/3, if they are four it is 0.25, and so forth. See, for example, Resnik (1987, pp. 35–37).

[16]Rawls (1971, pp. 167 ff.).

[17]After all, also Harsanyi's probabilities are derived with the help of empirical information (although not exactly the information Rawls would have preferred) and what probabilities are not in their essence statistical constructs?

[18]This in contrast to the (unsupported) claim of, e.g., Allhoff (2009), that PP should be assumed to be a supplement to traditional risk analysis that applies only in cases of "uncertainty".

[19]Gärdenfors' and Sahlin's general idea with regard to how the notion of epistemic risk may be applied via a preference for robust probability estimates seems to run along such lines, see Gärdenfors and Sahlin (1982).

the contrary, the basic version of the standard view (where values are determined by preferences) would seem to reject any such notion of rational action with regard to a choosing party who does not care for precaution. In other words, the support of this type of move would necessitate other considerations than those of (instrumental) rationality to be taken into account.

This brings us to the third and final objection to the idea of justifying the requirement of precaution on the basis of the standard view of rational action – and thus to identify the ideal of the desirability of precaution with the maximin rule. This objection connects to the former one but also to the discussion above about the actual *normative status* of the standard view. As was explained in connection to the criticism of the enlightment ideal for being instrumental, the standard view of rational action does not really have any ambition of issuing categorical prescriptions regarding what decisions should be made and what actions should be performed. What the standard view says is merely that this or that decision or action is *rational*. But, as we saw above, that claim may be true although the decision in question should actually not be made, all things considered, and that the action under consideration should, on the whole, not be performed. In effect, if the requirement of precaution would be interpreted as a principle of rational action under conditions of ignorance in the form of the maximin rule, it would be quite compatible with a principle stating categorically that certain risks should not be taken even if they can be described in numerical terms and even if taking these risks would maximise expected utility. For example, even if the decision to opt for a very unlikely but extremely valuable gain at the cost of almost certain horrible suffering being effected on a large number of people would maximise expected utility (due to the enormity of the possible gain), it may still be claimed that such a decision, in spite of being rational, should not be made all things considered.[20]

In all, therefore, interpreting the requirement of precaution within the framework of the standard view of rational action seems to take us only so far. What we get is a far from obvious idea that is crippled by the condition of ignorance and which leaves plenty of room for other normative reasons to counteract and override its prescriptions. However, as we now will see, already in the context of the discussion regarding the maximin rule described above, we find openings for moving our quest beyond rationality and base our interpretation of the requirement of precaution on more powerful normative considerations.

---

[20] In Sandin (2004, pp. 7–8), a related conclusion regarding the relevance of decision theory to the interpretation of PP is presented by pointing out that PP (or the requirement of precaution) may be seen as a "transformative" decision rule. This is the idea that decision theories may incorporate the idea of 'meta-rules' restricting what options are permitted into the final calculus of rational decisions (so that, for examples, options bringing risks of catastrophic outcomes are excluded from the final analysis in spite of the fact that they might maximise expected utility). My point in connection to this is that the *justification* of such rules, if they are to be fitting for the task of illuminating PP, needs a theoretical underpinning that clearly goes beyond the standard view of rational action.

3.3 From Rationality to Morality

As described above, the general argument for precaution within the framework of the standard view of rational action purports to establish, within the traditional assumptions of decision theory, the rationality of deciding according to the maximin rule in some special circumstances. However, it is an interesting fact that Rawls himself, the most eloquent spokesman of this view, actually puts forward a *further* argument which has much more of a moral ring to it.

3.3.1 Rawls' Appeal to Responsibility

In his response to Harsanyi's suggestion of basing calculations of expected utility on 'lottery probabilities', Rawls claims not only that acting on such calculations would be *unwise*. He also claims that even if this had been rational, such decisions would still be viewed by most people as plain *irresponsible* and therefore be seen by them as *categorically* unacceptable.[21]

The relevance of this for Rawls is the interest of the parties, who are to choose the basic constitutional structure of society in his theory, that they make a choice that can be accepted by most members of the society resulting from their choice. Therefore, he does not himself put forward any substantial claim regarding what choices would actually *be* irresponsible. However, it still seems that Rawls here goes beyond opinions about rational action within the framework of the standard view and refers to opinions that have a definite moral flavour. For, the consequence of the assumption that people have ideas about the responsibility of choices in the way described by Rawls seems to be that they may consistently hold that although someone's choice was *rational*, it was nevertheless *irresponsible* and therefore faulty in a much more fundamental way than suffering from the lack of instrumental efficiency making up the core of the standard view of rational action.

Of course, in light of the objections to the case that can be made for precaution within the standard view of rational action, this opening to more categorical normative reasons for precaution is highly interesting. Moreover, even if Rawls only assumes people to hold convictions about the (moral) irresponsibility of decisions in connection to conditions of ignorance, such views do not, in fact, have to be restricted to such special circumstances. That is, ideas about the (moral) responsibility of decisions to take or impose risks may apply no matter if we consider decisions under conditions of risk, uncertainty or ignorance. From the point of view of the objections made above, this seems to be a special virtue of the idea of trying to base the justification and interpretation of the requirement of precaution on this kind of convictions. It does not matter here whether or not Rawls himself actually

[21] Rawls (1971, p. 169).

accepts or holds such ideas (his argument against Harsanyi certainly does not presuppose anything in that vein). The point is that he makes use of the apparent fact that it is possible to have such ideas and that he believes people to actually entertain them.

3.3.2 Moral Opinions About Risk Impositions

Now, this suggested line of further inquiry may seem to be undercut by the possible objection that the ideas about irresponsible decisions alluded to by Rawls are not really about risk taking and risk imposition as such. Rather, they can be reinterpreted as claims about actual outcomes or consequences: If people are dissatisfied with the outcome of the choice of the parties in the original position, they will complain that the decision making was irresponsible, that is all that has been shown. Or so it might be argued.

However, other cases can be described that are not open to such reinterpretation. Consider, for example, the following scenario:

> *My Former Neighbour the Terrorist*
> I am contacted by the national security service about my former neighbour, who has recently been detained on serious charges of terrorism. Learning that, for many years, my neighbour stored in his quarters substantial quantities of unstable explosive substances (e.g., nitroglycerine), I am outraged and loudly express the view that my neighbour's behaviour was unacceptable. Hearing this, the security officer raises his eyebrows in bewilderment and reminds me that the explosives are now in a safe location (as is my former neighbour), and that, since nothing actually happened, I have nothing to fear and, frankly, nothing to complain about. This, however, does not stop my outrage. Although I know that the former activities of my former neighbour never harmed me, I still find his behaviour highly blameworthy. Of course, this fault of his would have been even worse if I had in fact been harmed, but even discounting for this, since he nevertheless exposed me to intolerable dangers, I still have a valid complaint against his behaviour.[22]

Most people, I take it, would side with me on this. My neighbour's risky behaviour may very well have been rational according to the standard view, but there is still something wrong with it. The question is *what*?

Obviously, it will not do to say that the fault of my neighbour lies entirely in the annoyance he caused in me. For, in that case, it would seem that the security officer behaves no less wrong than my former neighbour did.[23] Neither will references to the malevolent and obviously faulty values of my neighbour regarding what is a desirable outcome suffice to justify my reaction. Although we may assume that my reaction indeed *presupposes* these values to be faulty (and although the fact that my neighbour holds these values may help explaining his

---

[22]Cf. Munthe (1999, pp. 161–162).

[23]Moreover, this annoyance is no worse than what I am exposed to overhearing someone making a racist remark to a friend. But, surely, my former neighbour's fault is much worse from a moral point of view than such an action.

risky behaviour), what I react against is *his production of a certain likelihood of certain events that are undesirable*. For this reaction to be justified, these events would indeed have to be undesirable, but that is not enough. It has to be the case that there is something morally problematic with merely *risking* such undesirable events.

Thus, I take it that Rawls was right in his hunch about us having strong moral ideas about what risks we should be allowed to impose on each other. This, however, is not the end of the story.

3.3.3 Moral Dilemmas of Precaution

If we look at the specific context of trying to make PP more precise, different clarifications of the requirement of precaution seem to make a substantial difference as to how we are to resolve hard moral problems, where facts about risk imposition have to be balanced against other clearly morally relevant facts. Therefore, not only does our choice of interpretation of the requirement of precaution awake what are clearly moral sentiments about the risks we create (e.g., in cases like that about my former neighbour), but it also has a bearing on how to resolve certain peculiar *moral dilemmas* – i.e., situations where we have to pay a morally significant price for avoiding the imposition of some risk.

This fact is most easily demonstrated in connection to the 'show' and 'too serious risk' dimensions described in the foregoing chapter. As I sketched there, depending on how we clarify the requirement of precaution in these dimensions, it will be more or less difficult or demanding to show that some activity meets its conditions and this, in turn, will affect the decision costs and thus what degree and price of precaution is prescribed.

Now, imagine a situation where we contemplate the use of genetically modified crop in order to reduce the serious environmental problems created by modern farming. This crop would not need the toxins and artificial fertilisers used with contemporary types of crop in order to produce a sufficiently rich harvest. However, scenarios can be described where the introduction of the new crop would in the future alter the ecosystem to such an extent that humans and many sentient animals would suffer very serious harm, and these scenarios are, we assume, sufficiently probable according to current scientific expertise for making the action to introduce the new crop into something that 'may bring great harm'.

In this situation, the requirement of precaution tells us that we should not use the new crop until it has been shown not to bring too serious risks. Furthermore, as was demonstrated in Chapter 2, the stronger the interpretation in the 'show' – and 'too serious risks' dimensions, the greater the difficulty of meeting this demand. In practice, what this means is that more time will be needed to collect the evidence required for 'showing' the possible fact that using the crop will not bring 'too serious risks'. But, of course, the more time being spent on this, the more time will also be allowed for the currently ongoing destruction of the environment caused by contemporary farming methods. And, since it will always be possible to expand and refine the evidence a bit further, there is in principle no end to the time that the requirement of precaution in this manner may prescribe environmental degradation to continue.

However, interpreting the requirement so strongly would, I take it, be seen by most people as moral lunacy. The protection against possible unwelcome future effects of the new crop provided by such a strong version of the requirement of precaution is simply not worth the price we have to pay in actual damage and harm. In other words, the price of precaution that such a version of the requirement would prescribe us to pay would be unacceptably high. However, at the same time, just starting to use the new crop on a large scale without taking any precautions whatsoever against future disaster would, most of us would judge, be just as preposterous. That is, in that case, the prescribed degree of precaution would be unacceptably low.[24]

This simplified example illustrates an issue that, for instance, seems to be of the highest importance for the ongoing debate with regard to political action in the face of climate change. Whatever measures are taken to prevent undesirable effects of this phenomenon, these will impose certain costs in terms of money, quality of life and possibly also life. At the same time, scenarios regarding the actual preventive effects of these actions have to be uncertain (partly since the basic climate change scenarios are uncertain). For this reason, we find opinions on both sides with regard to whether or not the price of different suggested precautionary measures is in fact too high (or too low, for that matter).[25]

So, where, then, do we draw the line? How much of actual environmental damage should we accept in order to investigate the possibility of future disaster being the effect of our putting a halt to our current damaging practice? More simple: what price in terms of certain present harm or value-loss are we permitted or obliged to pay in order to increase our security against possible future harm? Any more precise version of the requirement of precaution will need the support of arguments to the effect that its answer to the just posed query is morally acceptable.[26]

The upshot of all this seems to be that the requirement of precaution not only has the potential of being based on something more than shaky and normatively weak decision theoretical intuitions about rational action in the face of ignorance. Given the point made above, that we hold substantial moral opinions of relevance to the clarification of the requirement, we can now see that it is also crucial to explore further these kinds of opinions in order to resolve the type of moral dilemmas that has just been described. For if we do not, we seem unable to proceed any further in answering the key question regarding the proper price of precaution. In effect, there are very good reasons indeed for exploring the possibility of interpreting the requirement of precaution as based primarily on categorical moral claims rather than views about mere instrumental rationality.

---

[24]The general case of biotechnology is further discussed in Chapter 6.

[25]The case of climate change policy is further investigated in Chapter 6.

[26]This type of problem is a well-known theme in medical research ethics, where a recurring question is that of when a clinical trial should be said to have given sufficient ground regarding the efficacy and safety of a new drug or procedure for it drug to be introduced as a routine treatment for all patients. See, for example, Tännsjö (1994).

References

Allhoff F. "Risk, Precaution, and Emerging Technologies." *Studies in Ethics, Law, and Technology* 3 (2) (2009). doi: 10.2202./1941-6008.1078.

Gauthier D. *Morals by Agreement*. Oxford: Oxford University Press, 1986.

Gärdenfors P. and N.-E. Sahlin. "Unreliable Probabilities, Risk Taking, and Decision Making." *Synthese* 53 (1982): 361–86.

Habermas J. *The Theory of Communicative Action*. Cambridge: Polity Press, 1991.

Hansson S.O. "The Limits of Precaution." *Foundations of Science* 2 (1997): 293–306.

Harsanyi J.C. *Essays on Ethics, Social Behaviour, and Scientific Explanation*. Dordrecht: Reidel, 1976.

Hume D. *A Treatise of Human Nature*. 2nd edition, edited by L.A. Selby-Bigge and P.H. Nidditch. Oxford: Oxford University Press, 1978.

Kant I. *Groundwork of the Metaphysics of Morals*. Edited by M. Gregor. Cambridge: Cambridge University Press, 1998.

Malmnäs P.-E. *Foundations of Applicable Decision Theory*. Stockholm: Department of Philosophy, Stockholm University, 1999.

Malmnäs P.-E. "Towards a Mechanization of Real-Life Decisions." In *Logic and Philosophy of Science in Uppsala*, edited by D. Prawitz and D. Westerståhl. Dordrecht: Kluwer Academic Publishers, 1994.

Munthe C. "Ethical Aspects of Risk Decisions." In *Amalgam and Health – New Perspectives on Risks*, edited by V. Novakova. Stockholm: Swedish Council for Planning Coordination of Research, 1999.

Parfit D. *Reasons and Persons*, 2nd printing. Oxford: Clarendon Press, 1984.

Rawls J. *A Theory of Justice*. Oxford: Oxford University Press, 1971.

Resnik M.D. *Choices: An Introduction to Decision Theory*. Minneapolis & London: University of Minnesota Press, 1987.

Sandin P. *Better Safe than Sorry: Applying Philosophical Methods to the Debate on Risk and the Precautionary Principle*. Stockholm: Royal Institute of Technology, 2004.

Smith M. *The Moral Problem*. Oxford: Blackwell, 1994.

Tännsjö T. "The Morality of Clinical Research. A Case Study." *The Journal of Medicine and Philosophy* 19 (1994): 7–21.

Chapter 4
Ethics and Risks

The conclusion reached in the foregoing chapter, that it is desirable to base the requirement of precaution on a set of categorical moral convictions, moves our quest into the area of ethics. However, this in itself takes us only so far. For, bold conjectures to the contrary notwithstanding,[1] it is far from obvious what kind of ethical reasons could actually accomplish this. Moral claims and ethical theories come in all shapes and sizes, both in the form of competing claims or theories answering the same moral question, but also claims or theories constructed to answer quite different albeit equally moral queries. A basic issue therefore seems to be where claims about the moral responsibility of decisions to impose risks – like the ones used by Rawls or sentiments of the type awaken by the examples with my former neighbour the terrorist and the case of the genetically modified crop – are to be located in this complex scheme.

In the following, I will consider three basic possibilities, of which the last one comes in two competing versions. First, I will investigate whether or not any of those traditional ethical theories that state criteria of morally permissible and impermissible actions can suit as a basis for justifying the requirement of precaution. Answering this question in the negative, I will then briefly turn to the possibility of locating the normative basis of precaution within traditional ethical theories about virtue, again with sceptical results. Finally, I will consider more recent suggestions for expanding traditional ethical theories in order to make them cover also the kind of peculiar issues that need to be addressed and plausibly resolved in order to justify the requirement of precaution. The outcome of this scrutiny is ambiguous. On the one hand, for several reasons, trying to expand any existing ethical theory does not seem to be a very fruitful path. On the other, however, this mainly serves to underline the need for a more worked out theory on the ethics of risk. What we have learned is mainly that this need must be met on its own terms – not merely by constructing minor appendices to existing bodies of ethical thought.

[1] See, for example, Haller (2000), who claims there to be obvious and strong ethical reasons for accepting PP (although it is a bit unclear exactly *what* PP he refers to), but unfortunately, albeit typically, without describing what these reasons are supposed to be.

C. Munthe, *The Price of Precaution and the Ethics of Risk*, The International Library of Ethics, Law and Technology 6, DOI 10.1007/978-94-007-1330-7_4,

4.1 Traditional Criteria of Rightness

One, if not *the*, central issue of normative ethics concerns the moral rightness (or permissibility) and wrongness (or impermissibility) of actions. What features of an action makes it morally permissible? What features would instead have made it wrong? Trying to answer these questions, philosophers and other ethicists have developed competing families of ethical ideas – each of which in turn consisting of a considerable number of mutually incompatible members.

4.1.1 The Diversity of Normative Ethics

A traditional way of describing these families is the distinction between consequentialism and deontology. That is, between those ethical theories that in one or other way base the rightness and wrongness of actions on the value of their consequences and those theories that claim some types of actions to be right or wrong in themselves, 'whatever the consequences'.[2] Classic examples of consequentialist ethical theories are *ethical egoism*, according to which every agent should always act to the greatest benefit of him- or herself, and *utilitarianism*, according to which an action is morally right if, and only if, none of those actions that could have been performed in its place would have effected more total well-being. Equally classic examples of deontological ethical theories are so-called *natural law ethics*, giving a list of types of actions the intentional performance of which are considered absolutely impermissible (such as murder, lying etc.), and *rights-based ethics*, according to which each person has an equal right to have his or her interests respected.

In recent decades, this traditional way of classifying ethical ideas has come to be overrun in several respects. For example, many versions of rights-based ethics seem to base their moral assessments of actions solely on what consequences will follow. However, they still differ from consequentialist theories in that they deny that any bad consequence can be balanced by just any comparably good consequence. For example, according to classical utilitarianism, even the smallest interest of each of a very large number of people can be aggregated into a 'sum of interests' that is capable of morally counterbalancing the most horrible suffering of a single individual. However, most rights-based ethicists would deny that, although they continue to claim the wrongness of an act of this type to be based solely on an assessment of its consequences (for the right-holder). What separate them from the classical utilitarian are rather their views on what aggregations and comparisons of individual interests that are allowed to influence the final judgement regarding rightness and wrongness of actions that is inferred from an assessment of its consequences.[3]

In a similar manner, the weight given to intention and motive in typical natural law ethics can be seen as a way of regulating when and to what extent the

[2]Cf. Bennett (1966).

[3]See, for example, Kamm (1993) for interesting discussions and suggestions regarding this.

actual production of good or bad outcomes should influence the moral rightness and wrongness of actions – i.e. as elements employed within a typical consequentialist calculus. In fact, we can even go so far as describing utilitarianism – normally considered to be the typical case of anti-deontology – as a particular version of natural law ethics that categorically bans only one type of actions (i.e., the type of actions marked by the feature of having at least one alternative that would have led to better consequences).

For these reasons, rather than juxtaposition consequentialism against deontology, an increasingly popular trend in contemporary normative ethics is to find and discuss more narrow and well-defined features of ethical thought. Here are some examples: The idea that making something bad occur is more morally serious than allowing it to occur. The idea that the intentions or motives from which an action is performed have a bearing on its moral status. The idea that certain behaviours or outcomes are absolutely morally prohibited. The idea that, for some morally acceptable actions, it is not the case that it would be wrong to do something else instead, although that would make one's behaviour possess less of the feature that gives rise to moral acceptability. The idea that the only reason for an action being wrong is if it would harm someone. The idea that such harm brought about by an action can only be morally counterbalanced by the fact that this action would prevent an equally serious harm to someone else. The idea that if more people are harmed or benefitted by our actions, this increases the moral reasons against them or in their favour. The idea that not only effects on the well-being of sentient creatures but also effects on the abstract relations holding between these creatures (such as the degree of inequality) or on the non-sentient structures within which they reside (such as species or ecosystems) have relevance for the moral status of our actions. And so on.

This increased diversity and specificity of ethical thought means that positions developed in contemporary normative ethics are not so easy to classify in any simple scheme. Answers to questions like the ones just listed can be mixed in ways so many and with a level of complication so high that the very idea of a clear-cut scheme of classification of ethical theories looks considerably more problematic than only a few decades ago. For sure, many writers defend answers to some of the above questions that would fit well with your typical self-proclaimed deontologist or consequentialist. However, since deontology and consequentialism is not very easy to separate in the first place, this does not tell us much. This will also be my own creed when, in later chapters, I proceed to present my own suggestion for the ethical basis of PP, and many of the sub-questions just mentioned will then reappear in that context. I see no point, however, in forcing this discussion into a preconceived division of ethical theories between consequentialism and deontology – although I will note when arguments or suggestions appear that remind of ideas that in the history of ethics have often been sorted under one or the other of these headings.

But, it is then only natural to wonder, can *any* general claims be made regarding the application of classic normative ethical thought to the issue of the morality of precaution and, eventually, the normative basis of PP? As will now be seen, there is at least one such claim, and this claim also forms the basis of a principal problem that has to be tackled by anyone that wishes to base an ethics of risk on traditional normative ethical notions.

4.1.2 Factualism and the Silence on Risks

What unites all classic suggestions regarding how criteria of moral rightness and wrongness should be constructed is that they have *nothing at all* to say about those moral sentiments and issues that are most relevant from the point of view of precaution. To be sure, they can say a lot about issues of some relevance to these things. They may, for example, help to clarify what makes for a harm, what makes harms more or less serious and to what extent the action of inflicting a particular harm can be morally justified or not. However, on issues regarding how we should morally assess the fact that an action *might,* or *is to some extent likely to* inflict such harm these same criteria all remain silent.

This rather striking feature has to do with a basic presupposition in the discussion regarding criteria of moral rightness and wrongness of what Erik Carlson has dubbed *factualism.*[4] If we look at the classic suggestions regarding what influences the moral rightness and wrongness of acts, these all operate on descriptions of situations with *given* actions and outcomes. That is, the very issue addressed by these suggestions is about how to evaluate morally our doings in terms of *actual* actions and outcomes, not the lesser or greater *likelihood* of some action being performed or some outcome being effected. For example, classic utilitarianism says that an action is morally wrong if, and only if, it actually maximises value.[5] Rights-based ethics claims actions to be morally acceptable if they do in fact not infringe any rights. Your typical list of forbidden types of actions in the spirit of natural law ethics communicates the message that actually performing any such action would be morally wrong. However, none of them says *anything* about the rightness or wrongness of *risking* going against the conditions of these criteria – i.e. the moral status of *more or less likely* maximising value, avoiding the infringement of rights or abstaining from actions of the types on the forbidden list. And, since this implication has to do with the basic presupposition of factualism, the same story repeats itself with more complicated ethical theories emanating from the classic suggestions just mentioned.

---

[4]Carlson (1995, pp. 20ff). The factualist claim formulated and discussed by Carlson only concerns utilitarianism. However, it can easily be expanded into a claim fitting any standard suggestion regarding the rightness and wrongness of actions.

[5]The 'founding fathers' of utilitarianism were not always entirely clear on this point. In particular, Jeremy Bentham (1907) expressed himself in a systematically ambiguous way as to the question if the important thing from a utilitarian point of view is that value is in fact maximised or if it *can be expected* to be maximised. Moreover, John Stuart Mill (1993) now and then expressed the spirit of utilitarianism as the idea that an action is right in proportion to its *tendency* to maximise value (which may but need not be interpreted in terms of the likelihood of this action to actually maximise value). Nevertheless, the standard interpretation of classic utilitarianism has come to disregard these ambiguities and settle for the factualist version.

Obviously, what this means is that no classic standard criterion of rightness coming out of this tradition of normative ethics can help us resolve the issue of the proper price of precaution. Since they cannot tell us anything about the moral seriousness of imposing risks, they cannot underpin moral convictions of the kind exemplified by Rawls' appeal to opinions about responsible decision making in the face of risks or the example of my former neighbour the terrorist. Even less can they tell us something about how the moral significance of risks should be balanced against other relevant factors, such as actual harms or other evils or certain losses of goods or failures to do what normally would be considered as one's duty. In effect, they cannot help us resolve those peculiar moral issues and dilemmas facing us if we want to make sense of the requirement of precaution. Having said this, however, there are two ways in which this claim might be disputed.

4.1.3 Autonomy and Justice

As was briefly mentioned in connection to the 'too serious' dimension in Chapter 2, suggestions have actually been made to the effect that at least some classic ideas of the rightness and wrongness of action, notwithstanding their adherence to factualism, may indeed be applied to questions about risks. The opening for such an application is created by postulating that risks can be seen as a special case of harms and burdens in general. Hence, risk impositions may be evaluated from criteria applicable only to the actual infliction of harm. For example, the basic idea of respect for autonomy (construed either as a basic right not to be subjected to coercion or manipulation, or as a value consisting in people not being subjected to such things) can in this way be directly applied to every case where one person imposes a risk on someone else. If the imposed risk is unwanted by the recipient, or in more complicated ways is in conflict with this person's plan of life, we can say that the imposition of this risk fails to respect this person's autonomy. Exactly how morally serious this is must then, in turn, be decided on the basis of a more comprehensive ethical theory stating to what extent autonomy restrictions may be morally justified. Although being most common within the framework of absolutist rights-based ethics,[6] this idea could presumably be applied to risk impositions in more or less complicated ways within the framework of all standard suggestions on what makes for a right- or wrongful action.[7]

Another example is provided by ethical theories of justice – i.e. theories stating what makes for a just or fair distribution of benefits and burdens. If risks are included among the burdens that may be distributed in different ways in a population (and

[6]McCarthy (1997), Nozick (1974, pp. 73–76), Shrader-Frechette (1991), Teuber (1990), and Thomson (1986).

[7]Thus, Hansson has suggested that a key issue for the ethics of risk is to develop a theory of a *prima facie* or *defeasible* right not to be exposed to risks (see Hansson 2003, 2009). See also Munthe (1999b, 1997b).

chances of actual benefits are included among the benefits), standard criteria of the acceptability of distributions of actual benefits and burdens may be applied also to the distribution of risks and chances in the population.[8] Again, it would seem that such applications would be possible no matter what specific theory of what makes for just distribution is considered (although, of course, the actual judgements regarding what actual distributions of risks and chances *are* in fact just and fair would presumably differ depending on what particular theory is applied).

However, although these kinds of ideas may have some bearing on how to interpret the requirement of precaution – i.e. the issue of what is an acceptable degree of precaution – they cannot by themselves resolve this issue.

First, they both rest on the basic assumption that risks and risk-impositions are indeed morally relevant categories in themselves. That is, in order to see patterns of the distribution of risks and people's wants regarding what risks they are exposed to as factors worthy of ethical consideration in their own right, we have to make sense of the basic idea of risks as comparable to actual harms or burdens. In order to illustrate this, we may consider the case where we are unsure of what effects our choices would have for how justly distributed and in tune with people's wants would be not only the actual harms and burdens created by our action, but also the *risks* of such harms and burdens. That is, even if we are equipped to assess actual impositions of risks of harm, we would still remain speechless when it comes to the question of how to assess choices in the face of risks regarding what actual risks would be imposed by our choices. In effect, both of the suggestions just sketched need a basic account of the moral relevance and seriousness of risk impositions capable of morally assessing risks not merely as a special case of harms and burdens, but in their own right.

Moreover, even if such an account is assumed to be available, there remains the basic issue of how risk impositions and actual harm inflictions should be balanced against each other in the moral assessment of situations where we have to choose between them. That is, besides a basic account of what makes for the moral seriousness of risk impositions, we also need a way of connecting this account to standard ideas of the moral assessment of actual harm inflictions. And, most important, this connection must be such that it makes possible moral comparisons of the seriousness of risks of harm and actual harms without compromising the underlying idea of risk impositions carrying a moral weight of their own.

It must be stressed that this conclusion does not amount to undermining completely the idea of approaching the ethics of risk from a theory of justice and/or autonomy. Rather, what I propose is that such applications of familiar ethical approaches need a theoretical complement in order to work, and it cannot be assumed from the outset that this complement will assess risks according to the standards of justice and/or autonomy employed in said approaches. And if they do not, we face the problem just mentioned of how to balance risk impositions and harm inflictions. Having said this, we will see in later chapters (especially Chapter 6) that

[8] See, for example, Perhac (1999) and Schuyt (1998) for attempts to this effect.

the theory that is advanced in this book actually *has* implications for that sort of distributive issues that typically occupy the attention of ideas about justice. Regarding autonomy, however, the result rather goes the other way, at least in relation to *rights-based* approaches to autonomy (this point is argued in Chapter 5). Still, autonomy may always enter as a proposed value that influences how bad the outcome aspect of a risk is and may furthermore be considered as a factor in the determination of the moral seriousness of risks, although this factor does not qualify as a moral right.[9]

4.1.4 The Two Level Approach

The problem of applying factualistic criteria of rightness and wrongness to cases where we do not know what choices will actually be right or wrong according to these criteria have indeed been the subject of quite a lot of philosophical reflection. Primarily among moral philosophers leaning towards one or other form of utilitarianism, the idea of *two levels* of ethical theories – one (factualistically) stating what makes actions right and wrong and the other describing how decisions should be made in practice – have been dwelled upon in various versions.

The roots of this idea can be found in Henry Sidgwick's famous claim that, from the point of view of utilitarianism, it may very well be the case that our practical moral thinking should be made along lines quite opposed to the ideas expressed by the utilitarian criterion of rightness. That is, it is far from sure that actually trying to find out what action would maximise utility and then try to perform the act one finds to do so is a method of decision making that would itself maximise utility. For this reason, utilitarianism may tell us to undertake our practical deliberations when deciding what to do according to some other principle than utilitarianism, albeit for reasons valid only in virtue of this ethical theory.[10]

In more recent times, Sidgwick's observation has been elaborated through the distinction between, on the one hand, ethical theories stating basic criteria of rightness and wrongness and, on the other, ethical principles serving as practically usable *decision making procedures* that should be applied in virtue of the fact that they are recommendable from a valid criterion of rightness (although they may, in particular situations, recommend actions that are wrong according to this very same criterion).[11] A further addition is the idea of nurturing behavioural dispositions, general traces of character and motivational patterns that would on the whole maximise utility, even though they may in particular instances lead us to perform wrongful

[9]This conclusion is a result of a combination of partial results in later chapters. The final rejection of the idea of a *right* against risk impositions being a basic notion in the ethics of risk is set out in the section on the weight of evil in Chapter 5 and summarily described at the end of the same chapter.

[10]Sidgwick (1907, Chapter 5).

[11]Bales (1971).

actions.[12] This division between two levels of moral thinking has then been perfected by adding the final touch of some room for almost straightforward utilitarian calculation within our decision procedures. In the spirit of the idea of rationality as calculated risk taking, we may sometimes try to do something close to trying to maximise *expected* utility (when the situation is right and time allows it).[13]

In spite of the fact that the development of the two level approach has been a utilitarian affair, its bearing thought seems to be equally applicable no matter what suggested factualistic criterion of rightness is considered. It holds for any such criterion that it tells us to be disposed to make our decisions according to a mix of practical rules and behavioural and motivational patterns that is recommendable from the criterion of rightness under consideration. And it is by no means given that the content of this mix is identical with the criterion of rightness itself. For example, if it would turn out that using your typical theory of natural law ethics as a practical decision tool would result in the performance of actions that are wrong according to this theory (perhaps due to some infantile psychological mechanism of compulsive rebellion against perceived authority) or if the actual fostering, nurturing and application of such a decision making procedure would violate the tenets of this theory, natural law ethics would recommend us to use some other decision making procedure instead.

In effect, the two level approach may seem to supply us with a generally valid strategy for justifying rules regarding how to approach and handling decisions where we do not know what action will be (factualistically) right or wrong. From the point of view of utilitarianism, we should do this according to a decision making procedure the application of which would actually maximise utility. From the point of view of natural law ethics, the recommendation is instead to use a procedure the application of which would respect the prescriptions of this theory. And so forth. In this way, we are supplied with a basis for saying that, in situations where we risk acting wrongly, we may proceed in a morally defensible way even if we, as it turns out, actually fail to avoid wrongful behaviour. Even if our action turns out to be wrong, we have still made our decision according to a procedure that is the morally recommended one. And, reversely, even if our action is morally right, our way of deciding to perform that action may be the object of a valid moral complaint if it is not the right way to approach risky decisions.

In consequence, it may appear that the two level approach provides a way of using factualistic criteria of rightness and wrongness for morally justifying some version of the requirement of precaution in the form of a decision making procedure. Or, at least, some idea regarding what degree of precaution should be exercised in the face of risks and chances can be so justified. Simply put, we should exercise that degree of precaution – thereby being prepared to accept those decision costs and pay that price of precaution – resulting from the application of a decision making procedure which is justified on the basis of the valid criterion of moral rightness

[12] Hare (1981, Chapters 2 and 3).

[13] Tännsjö (1998, Chapter 2).

and wrongness (whatever that is). In this way, then, it may appear possible to base the requirement of precaution on a factualistic theory of right- and wrongful action, although such theories do not really say anything about what risks we are allowed to take and impose. In so far as we are able to show that it would be right to apply some kind of precautionary approach to decision making according to the theory in question, we thereby supply a solid moral justification of this version of the requirement of precaution.

However, the small slit of hope that seems to open thanks to the two level approach is, in my view, unfortunately more of a mirage. Hard as it may be to know what particular action is (factualistically) morally right when we lack the relevant information in a particular situation, to identify what manner of practical decision making would be right to apply as a *general approach to risky decisions* seems exponentially more difficult if this is to be done on the basis of a factualistic ethical theory. Applying any minimally plausible factualistic criterion of moral rightness and wrongness, we will never know what approach to risks will best meet this criterion (since risky situations are characterised by exactly that feature of lack of knowledge regarding what should be done according to a factualistic criterion of rightness). In addition, there is quite a lot at stake when choosing what decision making procedure to use, since the wrong choice may lead to catastrophes of enormous proportions, no matter what factualistic criterion of rightness we take as the basis for our choice. In effect, the choice of practical approach to risky decisions will itself be a risky decision of sorts, and one to which the requirement of precaution would seem to be clearly applicable. At least, we are in no position to assume the opposite. However, since – according to the suggestion at hand – we do not know what the requirement says before we have made our choice of approach, we cannot know what approach to choose. Thus, the two level approach cannot solve the problem of morally justifying any particular version of the requirement of precaution.[14]

This leaves us with one remaining possibility of providing a moral basis of precaution from ethical theories of right- and wrongful action. Such theories must be modified so that they actually say something about the taking and imposition of risks – i.e. they must abandon the basic presupposition of factualism. Before investigating this possibility, however, I will first briefly comment on (and dismiss) another possible strategy for making sense of and clarifying the requirement of precaution on the basis of classic standard ethical thoughts.

4.2 The Virtue of Precaution

Aside from the moral status of actions, ethics and moral philosophy have also concerned itself with the moral status of *people*. Even if you act wrongly now and then, it still seems to be an open question how *you as a person* transcending the

---

[14] A much more elaborated and in-depth critical investigation of the two-level approach can be found in Gren (2004, Chapter 4).

particularity of any action you perform should be assessed from a moral point of view. It is even quite possible to imagine that you, for various reasons, act wrongly very often, but nevertheless is a perfectly *decent person* – far from that wicked brute your actions might lead others to suspect you for being. In other words, when we think about the moral qualities of people in the sense of persons seen as complex wholes (not only particular performers of particular actions in particular situations), this is something else than any simple function of the right- or wrongness of their actions. The moral qualities of you as a person have more to do with your basic personality and psychological characteristics (such as cognitive, emotional, habitual, intellectual and motivational patterns). Moreover, since they are detached from the urgency, polarisation and practical necessity of coming to a clear-cut decision that typically is a feature of the ethical issues of actions, such qualities need not limit themselves to the binary evaluation in terms of 'right or wrong?' with regard to this or that personal characteristic. Rather, to be a morally good person has to do with how you are composed personally and psychologically *as a whole* – something our language supplies us with quite vague and ambiguous words to describe. Nevertheless, we have and use these words – notions like kindness, thoroughness, wickedness, stupidity, wisdom, courage, generosity, and so on – and we use them mostly to attribute evaluative qualities, often moral, to people.

This summarises the basic point of departure of so-called virtue ethics – the virtues being those characteristics that are present in a morally good (or *virtuous*) person. Within this ethical tradition, different suggestions as to what characteristics are to be counted as (the real) virtues and attempts at describing desirable mixes of the virtues are debated. Having so been at least since Plato had Socrates suggesting that the fully 'just' (in the sense of morally good) person is characterised by a peculiar mix of wisdom, courage and primitive appetites,[15] the most prominent exponent of virtue ethics is otherwise Aristotle, whose basic scheme for the analysis of virtues and vices continues to influence present day virtue ethics.[16] Perhaps, therefore, it should come as no surprise that it is within this scheme that it may appear possible to find a place for a moral basis of precaution.

To be more specific, the virtues in Aristotle's scheme are analysed in terms of the balancing between extremes – between excess and deficiency. In view of the preliminary analysis of the requirement of precaution undertaken in Chapter 2, this may seem congenial to our envisioned moral basis of this requirement, since we there saw that this basis must be constructed by avoiding extreme versions of precaution (as well as lack of precaution) and the determination of a degree and price of precaution that is neither too high, nor too low. In exactly this spirit, we also find among commentators of Aristotle suggestions to the effect that the Aristotelian virtue of

---

[15] Plato (2007).

[16] Aristotle (1980).

courage indeed includes an ideal of (pre)caution.[17] Moreover, in one case, this idea has also been related to the discussion on PP, however, with rather uncertain conclusions.[18] Among philosophers more inspired by Kantian ethics, similar suggestions regarding the virtuousness of precaution have been made.[19]

Interpreted in this way, the requirement of precaution says that, unless you are disposed to deliberate and/or choose according to the guidelines and limits set by the requirement, you fail to express the virtue of precaution (or the proper amount of precaution according to the virtue of courage).[20] Some specific interpretation of the requirement could perhaps then be supported on the basis of a more detailed specification of this virtue.

Admittedly, this line of thought connects to the very naming of PP, since being cautious or disposed to take precaution seems to denote behavioural/emotional dispositions or character-traits of people, as well as to generally valued personal characteristics such as being thoughtful, careful, responsible, meticulous etc. Also, this angle of looking at PP has been used in the debate, by discussing whether a responsible decision maker is to be characterised as an optimistic or risk-taking individual, a calculating risk-taker or risk-neutral individual, or a pessimistic or cautious individual.[21] That is, the discussion of PP has sometimes been framed in terms of categorisations of people that correspond quite well to Aristotle's descriptions of how the ingredients of courage may be balanced so as to produce excess, deficiency or virtue.

In spite of this, although one could perhaps make a case for (pre)caution as a virtue, I do not believe this to be a very fruitful path when it comes to the context of making sense of and justifying PP or the requirement of precaution. First, just appealing to general intuitions regarding the value of personal characteristics like foresight, thoughtfulness, meticulousness etc. seems to be insufficient for straightening out those key unclarities in the requirement of precaution that have been identified above. Even if we agree that a disposition for *some* degree of precaution is a desirable personal trait, it still remains to clarify exactly *what* degree of precaution is required for being a virtuous person. And, as have been demonstrated earlier, in order to do this we must be able to specify what price of precaution it is acceptable to be disposed to pay in various situations. Thus, even if we grant that precaution may be a virtue, we must still deal with the question of how to balance the avoidance

[17] See, for example, Pears (1980), Ross (1949, pp. 205–207), and Urmson (1980, pp. 169–170) for suggestions in this direction. Aristotle himself never used the notions caution or precaution, but rather spoke about the proper role of *fear* in the virtue of courage.

[18] Sandin (2004, pp. 12–13) and Sandin (2009).

[19] Axinn (1990). In Kantian ethics, the focus is not on the general features of a person, but rather on those particular motives (or, as Kant would say, 'maxims') from which people act (see Kant 1998). However, in the present context, this difference may be disregarded.

[20] Of course, this virtue cannot then be explained in terms of having the disposition to act in accordance with the requirement of precaution, but in some independent manner.

[21] See, e.g., Munthe (1997a, Chapter 5), and Hansson (1999).

of risks against actual costs and harms resulting from this in the context of particular decisions.[22]

Secondly and partly in consequence of this, in the context of PP, we seem to be dealing with an idea obviously directed at guiding decision making and action. Thus, the ideal of the desirability of precaution expressed by the requirement would seem to be more about the responsibility of actions and decisions than the responsibility of acting parties and decision *makers*. From the point of view of this requirement, decision makers may be as adventurously, carelessly and recklessly disposed as they like, as long as the decisions they make meet the conditions of the requirement.[23]

Thirdly, as has been stressed from the outset of this book, PP is first and foremost a *political* suggestion, which means that the requirement of precaution should be interpreted in a way that allows it to support such a principle. However, virtue ethical ideals seem extremely ill-suited for this task. At best, they can propose that societal decision makers should have certain dispositions or traits of character, or make a point of the production of some sort of social virtue as a central objective of society.[24] That, however, does not seem to take us very far in the complex political reality of typical market-economy oriented liberal democracies where political decisions are the products of a complex interplay between numerous individuals, institutions and organisations, all influenced by a host of differing and competing inclinations and societal forces. In effect, rather than concentrating on qualities of those parties influencing societal decision making, in our pursuit of an ideal of the desirability of precaution potent of underpinning justifications of precautionary *policies*, we must have access to a scheme of evaluating directly the *content* and *outcomes* of such policies.[25]

[22]The suggestion (see, for example, Sandin 2009) that these issues may be handled within virtue ethics by saying that we should handle risky choices in the way that a virtuous person would have done does not seem very helpful. First, it assumes that there is a clear-cut description of such people, while it seems that the issue of how to describe a virtue of precaution needs to be settled before we know what people are in fact virtuous and what, more exactly, makes it so. Second, if we have such a description, the reference to virtue may be dropped altogether, since we then have a criterion for determining how risky decisions should be handled that does not mention anything about virtue.

[23]Urmson (1980, pp. 164–166) makes this point in his analysis of Aristotle's doctrine of virtue as the mean between excess and deficiency and its relation to social ideals such as justice.

[24]The latter idea corresponds roughly to Rosalind Hursthouse's post-Humean account of a virtue-based theory of justice (Hursthouse 1991).

[25]The same goes, I would be willing to claim, for the sort of communitarian suggestions that are structured as a virtue ethics of politics. All such ideas will either have to face that any society is made up of many sub-societies, each nurturing and enforcing their own special communitarian values, or resort to the suggestion that all people should act in accordance with some common norm, regardless of what (sub-)cultural values they entertain. This is not to deny the basic idea of MacIntyre (1985), that what social policies are pursued may indeed influence the virtuousness of individuals. However, this does not imply that the question of what social policies are or should be pursued can be determined on the basis of a description of individual virtues and their relation to political structures. See further Chapter 6 below.

4.3 Abandoning Factualism

The next suggestion, then, would be that the requirement of precaution is a principle regarding the straightforward moral *rightness and wrongness of action*, competing with or complementing other ideas in this field. As was demonstrated above, in order to underpin such an understanding of the requirement, classic suggestions regarding what may make actions right or wrong will have to be altered or amended, so that they speak directly about the moral permissibility or impermissibility of imposing risks in various circumstances. That is, the basic presupposition of factualism – the idea that the right- and wrongness of actions have to be determined by the actual outcomes or types of these actions – has to be abandoned. Furthermore, some idea regarding how such risks should be balanced against actual events that are morally relevant would have to be presented.

Since the classic criteria of right- and wrongful action are built on the factualistic assumption, accepting some criterion that does not observe factualism will mean that the classic ethical theories are in fact rejected. Hence, if justified from such a criterion of rightness, the requirement will express norms that are in conflict with all classic (factualistic) ethical theories. This is a direct upshot of the fact that the kind of alteration here considered concerns the idea of introducing into ethical theory the idea of mere risks of harm or undesirable events as directly right- or wrong-making characteristics. In opposition to, for example, classic utilitarianism, the requirement will say that if a person performs an act that may bring great harm without showing that it does not bring too serious risks, he acts wrongly even if he in fact succeeds in maximising utility. Against rights-based ethical theories, the requirement will claim that even if no one's right is in fact infringed by some action (suppose, for example, that all affected parties consent to being exposed to the possible risks brought by the action), it may be wrong to perform it because of the seriousness of the risks it brings. And, finally, against traditional natural law ethics, the requirement will claim both that it may be our duty to act against 'the moral law' (for example, to break a promise) if that is necessary in order to act in accordance with the requirement, and that we may act wrongly even if we in fact avoid all actions prohibited by this law (due to our risking of such actions).

In the following subsections, I will consider two basic ideas to this effect. The first of these takes as its point of departure the structure of deontology – in particular, classic types of natural law ethics. From this basis, it tries to make a case for strict moral bans on the risking of certain types of upshots. The second idea instead proceeds from a typical utilitarian basis, and tries to introduce the idea of the utilitarian calculus quantifying over risks and chances either together with or rather than actual harms and benefits.

4.3.1 The Forbidden Risks Approach

As mentioned in passing in Chapter 2, in various discussions of the moral status of embryos and foetuses, officials of the Roman Catholic church (as well as other supporters of its doctrines) have argued that even though it may be the case that embryos or foetuses are not a human beings of the type that makes the alleged moral ban on murder applicable to their destruction, it may also be the case that they are. This means, the reasoning goes, that even if they are in fact not human beings the intentional destruction of which would be murder, it is (almost) as wrong to destroy these entities as it would have been had such destruction actually been murder. And this for the reason that it is (almost) as wrong to willingly take a risk of committing murder as to actually commit such a sin:

> From a moral point of view, this is certain: even if doubt existed concerning whether the fruit of conception is already a human person, it is objectively a grave sin to *dare to risk* murder.[26]

The structure and ultimate premise of this argument exposes one possible route to altering or amending traditional criteria of moral rightness so that they speak directly about risks of harm and wrongdoing. One simply expands the area of traditional bans within natural law ethics, so that it becomes prohibited not only to actually perform certain types of action, but also to *risk* such performance. For example, the argument against the destruction of embryos and foetuses just outlined expands the traditional message of the Fifth commandment, 'thou shallt not kill', into the more demanding requirement, 'thou shallt not kill or take a risk of killing'.

Now, ethical discussions about particular acts of murder and their applicability to various single cases may not seem very congenial to the context of PP and its implicit concentration on large scale effects in societal policy making. However, suggestions have been made that are more in line with this implied area of interest, while still expressing the same structural approach to the ethics of risks as the just mentioned idea of Catholic ethicists. This is the case with German theologian Hans Jonas' so-called *imperative of responsibility*, according to which we are never permitted to risk the extinction of the human species.[27] This suggestion is structurally identical to the earlier mentioned claims of Catholic ethicists that we are never permitted to risk the death of a human being in that it attempts at amending the Ten Commandments with the general idea that not only are we forbidden to produce certain results (such as a lie, a murder, an act of adultery etc.), we are also forbidden to risk such results. The difference is that Jonas' suggestion regarding what results we are forbidden to risk seems much more in tune with the context of PP in that it concentrates on scenarios clearly relevant to policy making (not least in the fields of technology and environmental action) and is arguably paradigmatic of large scale catastrophes.

How, then, would this kind of expansion of traditional ethics affect our view of the requirement of precaution? First, it clearly limits the requirement in the 'may bring great harm' dimension, so that precautionary action is required only in connection to activities that may result in the end of humanity. Moreover, since it strictly

---

[26] *Let Me Live*, p. 8. Several statements expressing more or less the same idea can be found in Mahoney (1984, p. 68).

[27] Jonas (1979).

forbids the risking of this kind of outcome, the required precaution would seem to be that, in order to be justified in pursuing some activity that may lead to such results, we must first prove that this activity does in fact not bring such risks. Since all concentration here is on the undesirability of this kind of scenario there is no reason available within the theory to limit what may be required of such a proof. Hence, the desirability of avoiding risks of the extinction of the human species is here given a very large weight and, in effect, this version of the requirement would prescribe a very high degree and price of precaution as soon as some activity is found to be within its range of applicability. Indeed, it would seem that the only clear limit set to this degree is that we are not to introduce additional risks of the extinction of humanity through our precautionary measures.

There are several weak points in this suggestion. One of these is that, since the forbidden risks approach in general and Jonas' idea in particular fail to take into consideration the degree of likelihood involved in risks, such ideas become open to the objection of systematically producing decisional paralysis. Since *any* act or policy *may* in some weak sense threat the existence of humanity to some extent (it may, for example, bring a very slight likelihood of human history becoming one generation shorter), it would seem that all activities (including the act of retaining *status quo*) in many situations would be strictly prohibited according to Jonas' imperative of responsibility. And the same can be said regarding the forbidden risks approach in general, no matter what type of outcome is awarded the status of being forbidden to risk.

In order to avoid such implications, this approach would have to be amended by some idea of *de minimis* risk, so that only likelihoods of a certain magnitude are capable of making up a forbidden risk. At the very least, some such idea would seem to be necessary within the 'may bring great harm' dimension of the requirement. However, it is quite difficult to see how any such limit could be determined on the basis of the general spirit of the forbidden risks approach. Since this approach is entirely preoccupied with the evil of those upshots that we are forbidden to risk, we seem to be forced to move into some other area of ethics if we are to find any basis for balancing this evil against likelihoods. Below, I will connect this weakness to a further objection and argue that, together, they suggest that a gradual, rather than absolute, moral framework provides a more fruitful starting point for expanding traditional criteria of moral rightness so that they become applicable to risk impositions.

It may also be specifically objected to Jonas' imperative of responsibility that concentrating only on the risk of ending humanity makes for an interpretation of the requirement that is much too narrow. Even if we discount for the anthropocentrism implicitly implied by this suggestion, most of us are able to describe scenarios regarding the fate of humanity that do not involve its extinction, but that nevertheless qualify as catastrophes of the magnitude relevant from the point of view of PP and the requirement of precaution. For example, consider policies resulting in not the extinction of humanity, but in huge numbers of people being subjected to horrible suffering, held in slavery, et cetera. For that reason, it may be a tempting idea to make the basic structure of the forbidden risks approach more sophisticated by introducing several types of outcomes that we are forbidden to risk. However, since this would imply the possibility of conflicts (i.e., in order to avoid risking one such outcome we might be forced to risk another), such an extended version of the idea of forbidden risks would also have to involve some kind of ranking of the relative importance of avoiding risks of different outcomes. For example, Jonas' suggested outcome may still be at the top of such a ranking, but other types of outcomes may be added underneath in a serial order, so that we may never risk an outcome on the list unless it is necessary in order to avoid the risking some outcome residing higher up in the ranking.[28] While such a sophistication of the idea of forbidden risks would presumably make the picture of what degree and price of precaution is prescribed by the requirement more complex, we would nevertheless retain the basic idea that it is worth paying very high prices for avoiding risks of the outcomes that appear on the list of forbidden risks.

However, the situation of the forbidden risks approach is even more problematic. For if this idea is to be seen as an amendment to traditional natural law ethics, we must also take into consideration its relation to the other parts of such an ethical system. This is due to the fact that the activities undertaken to avoid risks on the forbidden list may involve actions that are forbidden by other parts of such an ethical system. Unfortunately, however, none of the suggestions outlined above give any clue as to how we should handle conflicts between the requirement of precaution interpreted in this way (as an absolute prohibition on the risking of certain kinds of upshots) and *other* rules prohibiting the actual (rather than possible) undertaking of certain actions (or, if you like, the actual production of certain upshots regardless of whether or not there was a substantial risk of producing them). Suppose, for example, that the existence of disabled people very slightly diminishes the chances of the human species to continue to exist a little bit farther off into the future. This seems to imply that Jonas' imperative of responsibility in this case orders us to commit mass murder of disabled people – which, of course, is prohibited by the commandment 'thou shallt not kill'. Or, to take a less drastic example, suppose that a politician has made a promise to his electorate to implement certain policies of great benefit to them, but later learns (through the information of experts) that these policies will bring additional risks of humankind having a slightly shorter history than what would otherwise have been the case. Is he then allowed to break his promise in order to avoid this additional risk?

It may be thought that these problems are merely about working out more clearly the proper priorities within such an ethical system. However, on further inspection it transpires that the forbidden risks approach here faces a principal difficulty having to do with the basic structure of any kind of natural law ethics. This is the general problem of how to make priorities between sins that are each considered to be an infinite evil (and thus worthy of absolute prohibition). In traditional (well-structured) systems of natural law ethics, this problem does not occur due to the fact

[28] Such a suggestion has been sketched by Dag Prawitz (1980) in a Swedish debate on the ethical basis for evaluating the use of nuclear power.

that the sins are defined in such a way that the absolute bans of the system cannot conflict. However, this would seem to be impossible if we want to amend such a system with the forbidden risks approach (since the avoidance of forbidden actions may always force us to produce forbidden risks). The only viable solution to this problem would, it seems to me, consist in rejecting the very idea of absolute bans and thus move away from the structure of a natural law ethics and into a structure of degrees of evil, which allows trade-offs between different values and considerations in a less rigid way.

A classic strategy within deontology to avoid such a result is to insist on the moral relevance of a distinction between doing and allowing evil. In the tradition of rights-based ethics, the counterpart is the distinction between making and allowing harm to occur, which underlies the so-called no-harm principle. A paradigmatic example of the role that this distinction is supposed to play is the idea that while the moral ban on murder in natural law ethics rules out any action of intentional killing, it allows you to abstain from actions that would prevent such killing or people dying as an intended result of some activity. Similarly, while the no-harm principle provides a reason against inflicting harm on people, it does not provide a similar reason against abstaining from actions that would prevent or limit harm.[29] Thus, many potential conflicts between deontological prescriptions can be avoided as long as the theory reserves the main bans for "active" wrongdoing or harm-infliction.

The very idea of such a distinction, as well as its alleged moral relevance, is controversial to say the least.[30] However, even disregarding that, the problems involved in applying this notion in the context of an ethics of risk capable of making sense of the requirement of precaution in a way that may underpin PP seem insurmountable. The simple reason is that the sort of wrongdoing that PP and the requirement of precaution tries to specify is mainly (if not wholly) about *allowing* serious risks to ensue when this could have been prevented or made less serious by taking active measures. To be sure, PP and the requirement will have things to say about the creation or sustainment of risks as well. But, once again, what these notions have to say will mainly be directed at acting parties capable of preventing such risks (or discontinuing their imposition). The faults that may be pinned on these parties will, in effect, all be "sins of omission". For this reason, then, the idea of a moral distinction between doing and allowing evil (or between making and allowing harm) will have nothing to contribute to an ethics of risk of the sort needed for the justification of a more precise version of PP.

The other classic strategy for resolving the sort of internal conflicts created by expanding natural law ethics to cover also risk impositions is to focus on the intentions or motives of the acting party as an ethically decisive factor. I will discuss this idea further in the next chapter with sceptical result.

---

[29]The idea of basing the ethics of risk on some version of the no-harm principle has recently been suggested by Zandvoort (2009).

[30]See, e.g., Bennett (1995), Kagan (1989), and Munthe (1999a).

It may, perhaps, be possible to retain the general spirit of the forbidden risks approach in so far as we may continue to claim a very high degree and price of precaution regarding the risking of certain types of upshots. However, if the objections described earlier are to be avoided, this degree and price must not be entirely without limit. Supporters of the forbidden risks approach may, for example, claim that we are not allowed to perform forbidden actions in order to avoid forbidden risks. Or, perhaps, a more sophisticated hierarchy of moral prohibitions could be introduced, where bans on actions and bans on the risking of certain upshots are mixed in the ranking structure, so that certain otherwise prohibited actions may be allowed in order to avoid the risking of certain upshots (you may, for example, be allowed to utter a minor lie in order to avoid the very likely, immediate and extremely painful extinction of humanity). However, we are still required to create as well as retain as much of harms and risks that are necessary to avoid risks that are higher up in this supposed hierarchy – and this no matter the magnitudes of these former harms and risks. In this way, irresolvable conflicts between different parts of a system of natural law ethics amended by the forbidden risks approach might be avoided. The basic idea that some outcomes are such that we are never allowed to risk them no matter the price is thus retained, but with the small provision 'unless it is strictly necessary for avoiding the breach of an even more important absolute ban'.

This takes us to my last and, in my view, most devastating objection to the forbidden risks approach. Since this strategy of simple prohibitions on the risking of certain types of upshots does not take into account either the likelihood-aspect of the risk or the fact that other upshots than the ones we are forbidden to risk may be almost as bad, this type of theoretical construct seems to lead to clearly implausible results even if we disregard the possibility of conflict with other rules. Assume, for example, that we have to choose between a 99% likelihood of extinguishing 99% of humankind and a 0.0000000000002% likelihood of extinguishing 100% of humankind.[31] Here, Jonas' imperative of responsibility provides us with the morally detestable recommendation to go for an almost certified extinction of almost all human beings, although the risk of extinguishing all of humanity thereby avoided is so small that it is almost inconceivable. And – unfortunately – such results of the forbidden risks approach will keep coming as long as we keep to the idea of absolute prohibitions on the risking of certain kinds of upshots, thus disregarding degrees of likelihood as well as degrees of value-loss associated with other possible outcomes. Furthermore, even if we moderate these possibilities by introducing some version of the *de minimis* idea and sophisticated rankings of forbidden actions and risks according to what was sketched above, we would still get results of this type, since the basic idea of the forbidden risks approach is that we should be prepared to pay *very* high prices for the avoidance of risks on the forbidden list as long as these payments are made with the right 'currency'. The only way to avoid such upshots would

[31] In order to avoid Parfit's claim (1984, p. 453) that, because of the loss of future generations, extinguishing 100% of humanity is much worse than extinguishing 99%, these percentages may be seen as regarding the last generation of human beings.

be to reject the very basis of the forbidden risks approach and base the determination of the proper degree of precaution on a gradual balancing of possible harms and benefits and the respective likelihoods of these.

This last remark clearly connects to the objection made initially above, that the forbidden risks approach suffers from a principal lack of basis for comparing and morally assessing differing likelihoods of undesirable events. What we have now added is a similar inability to consider seriously not only differences of likelihoods but also differences of the undesirability of outcomes balanced by their respective likelihoods. Together, these two objections strongly suggest that in order to pursue a more plausible way of expanding traditional criteria of moral rightness into an ethics of risk, we need to move into a framework that allows for the type of comparisons that the forbidden risks approach seems unable to account for.

4.3.2 Trading Off Risks and Harms 1: Apples and Oranges

How, then, would we fare if these aspects where taken into account, thus leaving the simplistic format of absolute prohibitions? Douglas Lackey has argued quite insistently that if we allow more of trade-offs both between risks and actual outcomes, and between the various risk impositions that we have to choose among (taking into account degrees of both likelihood and undesirability of outcomes), sense can be made of the intuitive idea to "put the infliction of risk into the same moral ballpark as the infliction of harm".[32]

As an illustration, Lackey urges us to compare the following two alternative courses of action, between which we are assumed to be forced to make a choice:

1. Kill 2 people at random from a group of 200.
2. Expose each of 200 people to a 2% probability of being killed.[33]

He then claims that, although many people as well as most traditional moral theories would judge option no. 2 to be the right choice, it would in fact not be absurd to claim the opposite. If we formulate a theory capable of comparing morally the badness of the actual outcome described in 1 and the combination of likelihood and badness of the possible outcome described in 2, we may reasonably claim that the latter scenario is in fact the morally inferior one. Such a theory would involve the idea of assessing the degree of badness of possible outcomes weighted by the degree of likelihood of their actual occurrence on the same moral scale as the one on which we evaluate the badness of actual events. On such a theory, it might turn out that the risk produced by option no. 2 is worse than the actual outcome of option no. 1, due to the fact that the former involves a sufficiently serious possible harm combined

---

[32]Lackey (1986, p. 636).

[33]Lackey (1986, p. 634).

with a sufficiently high probability of this harm actually occurring. And this in spite of the fact that this risk – as all risks – may in fact never be actualised at all.[34]

Obviously, such a theory carries with it the prospect of solving the ethical issue set out earlier, regarding how much of actual harm we are allowed/obliged to accept in order to secure ourselves against imposing risks. Moreover, as was made clear by the weaknesses of the forbidden risks approach, this gradualist way of approaching the idea of an ethics of risk may seem to supply us with the tools needed for avoiding the requirement of precaution to prescribe unacceptably high degrees and prices of precaution. However, as will now be seen, there are still serious difficulties involved in making sense of the idea of trading off risks of harm against actual harms.

An initial reaction to Lackey's idea is that it will require a theory that compares apples and oranges. Assume, for example, that we apply Lackey's suggestion in a utilitarian framework, adding to its implied theory of value the idea that being exposed to risks is bad for people. The resulting theory then would have to tell us how we are to compare actual inflictions of harm with mere risks of such inflictions. For example, if I have to choose between giving John a rather painful kick on the lower shin and imposing a risk on Jane, consisting of a 80% probability that John (unintentionally) makes a nasty remark about her in her presence, the theory will have to provide an answer as to which of these consequences is the worst one.

My initial view of this is that, although both these consequences can perhaps be evaluated in terms of badness, these 'badnesses' still fall in quite different categories. Being bad is for a harm something rather different than what being bad is for a risk and, furthermore, this difference is not possible to quantify as being one of degree on a single scale of (morally relevant) value. The reason is that actual harms are events in time and space, while risks do not seem to be identifiable events in space-time in the same way (since for any risk it holds that it may never be actualised). When a person is harmed she suffers an actual distress of some sort (such as pain or frustration) that carries the normative importance of harm (simply put, she is made worse off[35]), while a risk of harm does not necessarily involve this (again due to the nature of risks that they might never be actualised). This, of course, should not be taken to mean that exposing someone to a risk of harm carries *no* normative significance. However, this significance must be seen as based on something else than the normative significance of people suffering harm.

None of this is to deny that both riskiness and harmfulness could be seen as *properties of actual events*, namely if probabilities or likelihoods are such properties. For example, the performance of an action *a,* which is an actual event, may have both the property of being harmful and the property of probably producing

[34]Carlson (1995, pp. 20–24) points out that this general idea is unclear until we specify whether the risks are constructs of objective or subjective probabilities or likelihoods (cf. the unclarity regarding this mentioned in Chapter 2). He also presents some arguments regarding this issue. Since the arguments I present below are not dependent on any particular assumption in this respect, I will disregard this complication in the present context.

[35]What it is that may make affected parties worse off (and thus qualify as harm) is in turn dependent on the choice of theory of value mentioned at the outset of Chapter 2.

harm (and thus the property of being risky). In spite of the fact that my own inclination is rather to view probabilities and likelihoods as properties of *descriptions* of events, I cannot think of any knock-down argument for preferring that view of the matter.[36] However, once again, although riskiness may be a property of *a* as much as harmfulness, what makes *a* have these two properties are quiet different things. For, while the harmfulness of *a* is partly constituted by *another* event, the riskiness of *a* is not constituted by (or a property of) any other event at all – *a* may bring a risk of harm although no harm actually occurs. That, I would say, is the whole point of having a concept of risk besides a concept of actual harm. Furthermore, this point explains why we have reason to resist introducing into normative ethical theories of the traditional kind, that deal exclusively with actual harms, risks as a further kind of right- and wrong-making characteristic.

Lackey, however, flatly denies this view of the matter and claims that "what is of moral importance in what we ordinarily call 'the infliction of harm' is itself nothing other than the infliction of a risk",[37] and then suggests a way of comparing various risks very much reminding of an expected utility scheme.[38] In other words, Lackey here chooses the opposite route to the suggestions mentioned earlier of seeing risks of harm as a special case of actual harms. Rather, according to Lackey, actual harms are in fact nothing more than special cases of risks and may therefore be compared with other risks on an evaluative scale constructed for the purpose of comparing the values of risks. This puts him in the same camp as J.J.C. Smart, Jonathan Bennett and a few others, according to whom utilitarianism should be formulated as a principle of maximising *expected* rather than actual (morally relevant) value.[39] And,

---

[36]Perhaps, the best argument would be that, although an event may imply or give rise to a certain probability that some description of the world is true (my throwing the stone in your direction makes it probable that it will hit you), this probability is not best seen as a property of that event. Rather, it is the *implication or causation of* this probability that is a property of the event. However, I cannot see that anything of importance in the present discussion hinges on this, so if someone wants to have it the other way I will not attempt any further persuasion.

[37]Lackey (1986, p. 636).

[38]Lackey (1986, p. 637, note 8).

[39]Bennett (1995, Chapter 3, p. 144), Gruzalski (1981), Jackson (1991), Oddie and Menzies (1992), Oddie and Milne (1991), Smart (1973, p. 47). Carlson (1995, pp. 20–24) describes some variations of this basic idea under the heading of *probabilism* and ends up rejecting them all in favour of factualism. However, this rejection is based on, first, the undefended presupposition that the two level approach can help us answer all interesting moral questions we may have in the face of risky decisions and, second, a similarly undefended assumption that the actual upshots of the actions we may ever perform are never in a morally interesting sense, unknowable to us. I have argued in favour of the rejection of the first of these assumptions earlier in this chapter (and, to be fair, in footnote 43, p. 23, Carlson admits that it may present a problem – although he also expresses doubt regarding the idea of ascribing some kind of intrinsic value or moral importance as such to risks in footnote 45, p. 24). The second assumption, as we will soon see, is instead rejected by one of the primary defenders of a probabilistic version of utilitarianism, Jonathan Bennett, who uses this as a reason for his acceptance of probabilism. In fact, Carlson himself actually states that he also would be inclined to accept probabilism if a claim of the sort defended by Bennett should turn out to be warranted (p. 20, n. 38).

presumably, according to Lackey, other ethical theories would do best to revise their content in a parallel manner.[40] In short, then, the problem of comparing apples and oranges is, according to Lackey, a chimera, since all that we need to compare are various types of oranges.

Lackey tries to support this idea with the claim that...

> [...any] infliction of harm can be decomposed into some basic action, not by itself the infliction of harm, and certain causal or perhaps conceptual consequences that constitute the harm. Since the basic action cannot suffice to produce the causal consequences, all it does is increase the probability that the harm will ensue.[41]

And, on the basis of this, he argues that...

> ...all our moral attention must center on the basic act, evaluated in terms of the risk it generates.[42]

This, however, strikes me as a poor reason.[43] Of course, Lackey is right in claiming that acts do not give rise to events in a vacuum, but only given a background of certain circumstances, present and forthcoming. But this goes for *all* events, not only acts, and, surely, we do not want to say that events never give rise to other events. Rather, we should look at statements about this as implicitly implying a certain given framework of yet other events taking place independently of the causing event, yet still influencing the occurrence of the caused event. The important thing in the moral case (i.e. where the causing event is an action) is that we count as relevant consequences only those events that the agent could have prevented (given the other events making up the circumstances) by acting differently. So, given the surrounding circumstances, flipping my hand forward in your direction while releasing the grip of my fingers around the stone I held firmly just a tenth of a second ago *will* produce the event that you are hit by the stone. Why? Because the stone in fact hits you, and it is true that had I kept my grip, not flipped my hand and so on, the stone would not have hit you – given the surrounding circumstances. Of course, *these circumstances* might also have been different in a way influencing the outcome (a higher atmospheric density or a strong headwind might for example have prevented the stone from reaching you). However, since this lies outside of my control, it is

---

[40]For classical rights-based ethical theories this would mean that immorality consists in not the violation of rights, but the probable violation of rights. For natural law ethical theories, it would mean that what is wrong is not the performance of actions that actually are of a type prohibited by the natural law, but rather actions that probably are of such a type.

[41]Lackey (1986, p. 637).

[42]Lackey (1986, p. 637). He also argues that a parallel argument can be devised for those *conceptual* consequences of acts that determine the type of an act relative to a deontological system of rules quantifying over generic actions.

[43]Here is a further, although weaker, argument to this effect: The kind of basic acts Lackey speaks about are 'tiny' behaviours such as a minimal movement of an arm etc. But such acts are not what we choose between in our moral deliberations – or evaluate in our moral appraisals of people's behaviour. In effect, a moral theory constructed to quantify *only* over basic acts will lack all connection to the moral issues we actually are trying to solve with the help of moral theories. For an argument to this effect, but in a different context, see Munthe (1996).

of no relevance for judging my actual action (compared to available options) less wrong in the actual circumstances.[44]

In consequence, then, Lackey's suggestion fails to establish a plausible basis for trading off risks against actual harms, since he tries to accomplish this by claiming that actual harms are ontologically equivalent to risks of harm. However, this failure does not exclude the possibility of claiming that, although actual harms and risks of harm are ontologically distinct, what matters *morally* is in fact only the latter. That is, the apples and oranges problem could be solved by a *moral* argument to the effect that, in spite of the fact that apples and oranges are different things, we need only bother ourselves with the oranges.

4.3.3 Trading Off Risks and Harms 2: Improving Practical Guidance

One version of such a moral argument has been put forward by Bart Gruzalski. He claims that a general deficiency of all types of factualistic versions of utilitarianism is that they fail to deliver particular norms that can guide decision making and action in practice. We encountered this complaint earlier in this chapter, and saw how it may be generalised to all factualistic (and minimally plausible) ethical theories. We also saw how the two level approach fails to solve this problem. Gruzalski's suggestion is that the problem might be solved if we instead abandon factualism, and reformulate the ethical theory at hand (utilitarianism in this case) in probabilistic terms that connect to what may be foreseen. Instead of requiring us to maximise utility (where utility is determined by a morally sound theory of value), utilitarianism then prescribes the maximisation of *expected* utility, where the various expectations (which may or may not be described in terms of probabilities) express to what extent we are able to foresee what consequences will result from available options.[45]

There are two weak areas in this suggestion. First, the argument assumes that the ability to provide norms that in practice are able to guide action and decision making should be seen as a desideratum for a plausible ethical theory regarding the rightness and wrongness of action. It is, however, perfectly possible to hold the view of C.D. Broad that...

> [...we] can no more learn to act rightly by appealing to the ethical theory of right action than we can play golf well by appealing to the mathematical theory of the flight of the golf-ball.[46]

This position may seem at odds with the needs created by the quest of clarifying the requirement of precaution. However, it is not – at least not necessarily. What this

[44]For a more in-depth argument to this effect, but in a different context, see Munthe (1999a).

[45]Gruzalski (1981).

[46]Broad (1930, p. 285).

quest has illuminated so far is a need to have ethics saying something substantial and basic about risky actions and decisions quite independently of what it says about actions and consequences in factualistic terms. This, however, may be accomplished while still leaving open the question of to what extent we may use such an ethical position as a practical decision tool. That is an issue that I will return to in Chapters 5 and 6.

One may, of course, concede the idea that ethics should have some bearing on practical matters and contain some aspects that may be put to practical use. This, however, does not imply that an ethical theory cannot be plausible unless it is capable of producing particular norms that can be used as decision tools in a straightforward manner. In my own opinion, exactly what such a requirement of practicality means for ethics is undecided.[47]

Second, even if we grant Gruzalski the just mentioned assumption, it is far from clear that a version of utilitarianism along the lines sketched by Gruzalski would actually solve the problem. In short, even if the consequences to care about are the ones that are foreseeable, they may still be extremely difficult to actually foresee. This holds especially if the theory requires that such foresight is formulated in terms of probabilities exact enough to be fed into an expected utility calculus.[48] As a consequence, the type of requirement of practicality assumed by Gruzalski seems to provide no reason for why utilitarianism (and other ethical theories) should be reformulated in probabilistic terms.

4.3.4 Trading Off Risks and Harms 3: The Knowability Argument

Jonathan Bennett has suggested an argument somewhat reminding of Gruzalski's that satisfies the above claims of mine regarding the ontological difference between harms and risks, while still making a case for the idea that morality *should* focus on risks rather than actual harms. However, Bennett's point is not that this would make moral theories more practically useful. Rather, his claim is based on a purely moral idea, that we are to be held morally accountable only for such effects that can be foreseen.

The basis of Bennett's suggestion is the claim that even if we grant the point that acts give rise to specific effects given a background of actual circumstances, cases can be described where no one is in any position to *know* – even under ideal conditions – which these effects in fact are. As an illustration, he offers a case where the circumstances surrounding my acting are partly indeterministic, thus making the effects of my acting partly indetermined.[49] He then suggests that...

---

[47]For a recent discussion of this issue that also provides further references, see Gren (2004, Chapter 5).

[48]Cf. Gren (2004, Chapter 3), and Feldman (2006).

[49]This must not be misunderstood. It is, of course, still the case that had I acted differently, the effect would not have occurred. The indeterminism is thus placed among those events other than my act that contribute to the occurrence of the effect.

> . . .someone does not act wrongly at T unless this is made to be the case by facts that obtain at T – that is (1) facts that pertain to T and (2) facts that pertain to other times but are deterministically implied by the world's state at T.[50]

This, in turn, is based on the underlying idea that. . .

> . . .a wrong action must be one which a well-enough-informed bystander could in principle have advised against, condemned, or deplored at the time of acting; and an agent who has acted wrongly should have at least a theoretical chance of learning from the wrongness of his action, seeing how he could have done better and may do better in the future. The bystander's-comment and chance-of-learning desiderata both require that the marks of the behaviour's wrongness be laid down in the world at T; and that constitutes at least part of the reason why behaviour is not wrong unless the facts that make it so are registered in the world at the time of acting.[51]

But why is this? Bennett offers the explanation that we have reason to accept a 'knowability constraint', implying. . .

> . . .a concept of moral wrongness which requires that the relevant facts not merely be registered upon the world at the time of acting but be such as might be known and understood by human beings.[52]

And, according to Bennett,

> Because wrongness is tied to what can be known, it is tied to probability. The only way anything can be known at T_1 about what will contingently be the case at T_2 is through probabilities, including inevitabilities.[53]

Via a number of cases, this then prompts Bennett to suggest that *only* facts about the (knowable) probable consequences of an action can make this action morally wrong, since making wrongness dependent on the actual consequences would make the moral status of our actions a question of sheer luck.[54]

However, I sense a gap in this argument.

Although it could be challenged,[55] Bennett makes a reasonably convincing case for the knowability constraint, so let us grant that for the sake of the argument. Let us also grant that our chances of knowing the actual consequences of our actions are created entirely through our chances of knowing the probabilities of these consequences. Does this imply that only the latter kind of facts can contribute to the

---

[50] Bennett (1995, p. 50).

[51] Bennett (1995, p. 51).

[52] Bennett (1995, p. 52). As Bennett himself mentions (pp. 52ff.), such a constraint can be strengthened or weakened by invoking requirements on the knowability, for example that it has to regard the agent himself under some specific particular circumstances (rather than any human being under any circumstances, all human beings under all circumstances, some group of human beings under in some class of circumstances, et cetera).

[53] Bennett (1995, p. 52).

[54] Bennett (1995, pp. 57–58).

[55] For instance, Bennett's argument seems to build on certain controversial assumption with regard to what moral philosophers call *moral luck*. See, e.g., the overview of the debate regarding this notion, and the further references given there, in Nelkin (2008).

wrongness of actions? Not as far as I can see. As soon as they are knowable (through our knowledge of probabilities), the actual consequences of our actions are not cases of luck and can therefore still make such contributions. It would, it seems to me, also be quite strange not to allow for this – a claim that can be demonstrated through one of Bennett's own examples where he compares three cases[56]:

1. P's action caused a disaster.
2. P's action caused a disaster, and this fact was knowable when P acted.
3. It was (knowably) probable at the time of P's action that it would cause a disaster.

Only in case 3, Bennett claims, do we have a fact that may explain the wrongness of P's action. However, and this is my objection, would it not be more in tune with moral sense to say that, although 3 certainly pinpoints a morally relevant characteristic, this can also be said about 2. From a moral point of view, it is certainly *not* "altogether irrelevant"[57] whether or not the disaster actually occurs (given that this was knowable at the time of acting) as a result of P's action. This claim is compatible with the suggestion that, even if this consequence does in fact not obtain in case 3, P has committed a serious moral fault. However, an actual disaster makes this fault even worse – to claim otherwise would, in my view, be a case of sheer cynicism with regard to the victims suffering real and actual harm.

This last statement also sums up where I believe Bennett to go astray in his treatment of this issue. The knowability constraint takes care of ideas regarding fairness in the assignment of moral culpability to agents – i.e. the conjecture that a sound moral theory should not burden agents with too heavy moral responsibilities. However, when the step is taken from this to the 'only facts about probable consequences matter' view, the road is blocked for mixing this concern with an equally morally legitimate concern for the fate of the victims of these agents' actions. Even if the first of these concerns is taken to be one of the key sources of sound moral convictions, the same goes for the latter. Where there never any victims suffering actual harms or losses because of our choices, the basic moral question regarding what we should do would not present itself as very pressing at all.

4.3.5 Trading Off Risks and Harms 4: Back to Square One

My conclusion from all this is that we have no reason to reduce in our moral theories actual harms to mere risks of harm, neither ontologically nor morally. Armed with this insight, we can now reconsider the initial example of Lackey described above. On the basis of the arguments made above we can see how Lackey's original suggestion is based on an ambiguity bound to lead us astray on the matter of

[56]Bennett (1995, pp. 57–58).

[57]Bennett (1995, p. 58).

moral comparisons of harms and risks. What Lackey asked us to consider was the following options:

1. Kill 2 people at random from a group of 200.
2. Expose each of 200 people to a 2% probability of being killed.[58]

However, option no. 2 may in fact involve two quite different scenarios. The most straightforward of these is that the members of the group are subjected to a lottery with 200 tickets, of which four means certain death. Interpreted in this way, option no. 2 thus necessarily involves the actual killing of a number of people just as much as does option no. 1 and can therefore be looked upon as a case of actual harm-infliction, easily comparable to that in 1. And since option no. 2 interpreted in this way means that four people are killed, Lackey's basic intuition that this option is worse than option no. 1 (where only two people are killed) would seem to be warranted.

However, Lackey's claim to the contrary notwithstanding, this does not tell us anything about how to assess risks of harm *as such* in comparison with actual harms. What distinguishes risks of harm from actual harms is that the risked harms may in fact never occur – risks as opposed to actual events may remain mere possibilities. But if Lackey's example is interpreted according to the outline above, this difference is completely eradicated – what we are left with is a simple case of comparing two different inflictions of actual harm. In order to bypass this objection, option no. 1 would have to be compared with a very different interpretation of option no. 2, expressed by the following description:

3. Expose each of 200 people to a mechanism that is directed to each person in random order and rigged so as to fire a lethal missile with a 2% probability in each case where it is directed at someone. [59]

In this case, I conjecture, it is not as obvious anymore either that the affected parties perspective points to option no. 1 as being preferable to option no. 3, or that the treatment of the affected parties in options 1 and 3 can or should be placed "in the same moral ballpark" at all.

This, then, takes us back to square one where the idea to "put the infliction of risk into the same moral ballpark as the infliction of harm" faces the challenge of how to compare apples and oranges. Above, I concluded that what makes harms morally significant is something else than what makes risks so significant. In addition, we have just seen that we have no reason to accept any of the claims 'harms are nothing more than risks' or 'a sound moral theory should only concern itself with risks'.

---

[58]Lackey (1986, p. 634).

[59]Lackey's own specification of what the 2% risk may involve is "a randomizing device attached to a bomb" (1986, p. 634). However, this is consistent with both of the interpretations I have described, since the device may work either as a lottery (selecting 4 people at random to blow up), or as a device which exposes each person in turn to a 2% probability of being blown up.

Even less can we plausibly meet the challenge by treating harms as risks with a probability of 1 and apply some probabilistic decision rule such as the principle of maximising expected utility, since that would amount to counting the same factor twice.[60]

Taken together with the above mentioned reasons to reject the forbidden risks approach, what we now see is the failure of the strategy to abandon factualism in order to make room for moral evaluation of risks within a traditional normative ethical framework. However, the basic idea of risks as proper objects of moral concern in their own right still stands. This makes attractive the suggestion that we should take seriously the notion that actual harms and risks of harm have to be evaluated from wholly different notions of badness. Or, differently put, beside the categories of practical rationality, of virtuousness and viciousness, of goodness and badness of outcomes or events and of factualistic rightness and wrongness of actions, there is a further category of normatively relevant facts with regard to *decisions to impose risks*. So, in trying to find a normatively valid interpretation of the requirement of precaution, it is to this category we should now turn.

References

Aristotle, *Nicomachean Ethics*. Oxford: Oxford University Press, 1980.
Axinn S. "Moral Style." *Journal of Value Inquiry* 24 (2) (1990): 123–33.
Bales R.E. "Act-Utilitarianism: Account of Right-Making Characteristics or Decision-Making Procedure?" *American Philosophical Quarterly* 8 (3) (1971): 257–65.
Bennett J. *The Act Itself*. Oxford: Clarendon Press, 1995.
Bennett J. "Whatever the Consequences." *Analysis* 26 (1966): 83–102.
Bentham J. *An Introduction to the Principles of Morals and Legislation*. Oxford: Clarendon Press, 1907.
Broad C.D. *Five Types of Ethical Theory*. London: Routledge & Keegan Paul, 1930.
Carlson E. *Consequentialism Reconsidered*. Dordrecht: Kluwer Academic Publishers, 1995.
Feldman F. "Actual Utility, the Objection from Impracticality, and the Move to Expected Utility." *Philosophical Studies* 129 (1) (2006): 49–79.
Gren J. *Applying Utilitarianism: The Problem of Practical Action-Guidance*. Göteborg: Acta Universitatis Gothoburgensis, 2004.
Gruzalski B. "Foreseeable Consequence Utilitarianism." *Australasian Journal of Philosophy* 59 (2) (1981): 163–76.
Haller S. "A Prudential Argument for Precaution Under Uncertainty and High Risk." *Ethics and the Environment* 5 (2) (2000): 175–89.
Hansson S.O. "Adjusting Scientific Practice to the Precautionary Principle." *Human and Ecological Risk Assessment* 5 (5) (1999): 909–21.
Hansson S.O. "An Agenda for the Ethics of Risk." In *The Ethics of Technological Risk*, edited by L. Asveld and S. Roeser. London & Sterling: Earthscan, 2009.
Hansson S.O. "Ethical Criteria of Risk Acceptance." *Erkenntnis* 59 (2003): 291–309.
Hare R.M. *Moral Thinking*. Oxford: Clarendon Press, 1981.
Hursthouse R. "After Hume's Justice." *Proceedings of the Aristotelian Society* 91 (1991): 229–45.

[60]That is, first we count the probability of a particular harm occurring at the time of acting in order to measure the initial risk, then we count it once again when this harm has actually occurred in order to measure the actual harm treated as a risk.

Jackson F. "Decision-Theoretic Consequentialism and the Nearest and Dearest Objection." *Ethics* 101 (1991): 461–82.
Jonas H. *Das Prinzip Verantwortung. Versuch einer Ethik für die technologische Zivilisation.* Frankfurt am Main: Insel Verlag, 1979.
Kagan S. *The Limits of Morality*. Oxford: Oxford University Press, 1989.
Kamm F.M. *Morality, Mortality. Vol. I: Death and Whom to Save from It*. Oxford: Oxford University Press, 1993.
Kant I. *Groundwork of the Metaphysics of Morals*. Edited by M. Gregor. Cambridge: Cambridge University Press, 1998.
Lackey D.P. "Taking Risk Seriously." *Journal of Philosophy* 83 (11) (1986): 633–40.
MacIntyre A. *After Virtue: A Study in Moral Theory.* London: Duckworth, 1985.
Mahoney J. *Bioethics and Belief.* London: Sheed & Ward, 1984.
McCarthy D. "Rights, Explanation, and Risk." *Ethics* 107 (1997): 205–25.
Mill J.S. *Utilitarianism*. New York: Bantam Books, 1993.
Munthe C. "The Argument from Transfer." *Bioethics* 10 (1) (1996): 27–42.
Munthe C. *Etiska aspekter på jordbruk* (Ethical Aspects of Agriculture). Jönköping: Statens Jordbruksverk, 1997a.
Munthe C. "Teknikbeslut och ofrivilliga risker" (Technology Decisions and Unwanted Risks). *Filosofisk tidskrift* (2) (1997b): 7–22.
Munthe C. "The Morality of Interference." *Theoria* 65 (1) (1999a): 55–69.
Munthe C. "Ethical Aspects of Risk Decisions." In *Amalgam and Health – New Perspectives on Risks*, edited by V. Novakova. Stockholm: Swedish Council for Planning Coordination of Research, 1999b.
Nelkin D.K. "Moral Luck." *The Stanford Encyclopedia of Philosophy*, 2008. Available online: http://plato.stanford.edu/entries/moral-luck/.
Nozick R. *Anarchy, State and Utopia*. New York: Basic Books, 1974.
Oddie G. and P. Menzies. "An Objectivist Guide to Subjective Value." *Ethics* 102 (1992): 512–33.
Oddie G. and P. Milne. "Act and Value: Expectation and the Representability of Moral Theories." *Theoria* 57 (1991): 42–76.
Parfit D. *Reasons and Persons*, 2nd printing. Oxford: Clarendon Press, 1984.
Pears D. "Courage as a Mean." In *Essays on Aristotle's Ethics*, edited by A.O. Rorty. Berkeley & Los Angeles: University of California Press, 1980.
Perhac Jr. R.M. "Environmental Justice: The Issue of Disproportionality." *Environmental Ethics* 21 (1) (1999): 81–92.
Plato, *The Republic*. London: Penguin Classics, 2007.
Prawitz D. "Rationalitet och kärnkraft" (Rationality and Nuclear Power). *Filosofisk tidskrift* 1 (1) (1980): 1–14.
Ross W.D. *Aristotle*, 5th edition. London: Methuen, 1949.
Sandin P. "A New Virtue-Based Understanding of the Precautionary Principle." In *The Ethics of Protocells: Moral and Social Implications of Creating Life in the Laboratory*, edited by M.A. Bedau and E.C. Parke. Cambridge, MA: MIT Press, 2009.
Sandin P. *Better Safe than Sorry: Applying Philosophical Methods to the Debate on Risk and the Precautionary Principle.* Stockholm: Royal Institute of Technology, 2004.
Schuyt K. "The Sharing of Risks and the Risks of Sharing: Solidarity and Social Justice in the Welfare State." *Ethical theory and Moral Practice* 1 (3) (1998): 297–311.
Shrader-Frechette K. *Risk and Rationality. Philosophical Foundations for Populist Reforms.* Berkeley: University of California Press, 1991.
Sidgwick H. *The Methods of Ethics*, 7th edition. London: Macmillan, 1907.
Smart J.J.C. "An Outline of a System of Utilitarian Ethics." In *Utilitarianism. For and Against*, edited by J.J.C. Smart and B. Williams. Cambridge: Cambridge University Press, 1973.
Tännsjö T. *Hedonistic Utilitarianism*. Edinburgh: Edinburgh University Press, 1998.
Teuber A. "Justifying Risk." *Dædalus*: *Proceedings of the American Academy of Arts and Sciences* 119 (4) (1990): 235–54.
Thomson J.J. *Rights, Restitution and Risk.* Cambridge, MA: Harvard University Press, 1986.

Urmson J.O. "Aristotle's Doctrine of the Mean." In *Essays on Aristotle's Ethics*, edited by A.O. Rorty. Berkeley & Los Angeles: University of California Press, 1980.

Zandvoort H. "Requirements for Social Acceptability of Risk-Generating Technological Activities." In *The Ethics of Technological Risk*, edited by L. Asveld and S. Roeser. London: Earthscan, 2009.

Chapter 5
The Morality of Imposing Risks

Let me sum up the results so far. In the second chapter, two desiderata for a sound interpretation of the requirement of precaution were formulated. First, such an interpretation must not lead to decisional paralysis – at least not systematically so. Second, such an interpretation must at least implicitly imply reasons for its recommendations (thereby avoiding the issuing of arbitrary prescriptions). These desiderata alone exclude a number of possible interpretations of the requirement, for example, those employing simplistic conservatism or 'worship of the novel'[1] and those employing proof-standards that for most people are impossible to live up to in most cases. This leaves a number of possible interpretations prescribing rather different degrees and prices of precaution, in turn, effecting quite different demands on practical action. We also saw that, due to the holistic nature of precaution, variations with respect to such degrees and prices cannot be reduced to any single dimension of precaution, but are effected by variations in all these dimensions taken together. Thus, in addition to satisfying the above mentioned desiderata, a sound interpretation of the requirement of precaution must also be supported by normatively valid reasons showing that it prescribes a proper degree and price of precaution. Exactly *how* such normatively valid prescriptions are effected (i.e., what variations of the dimensions of precaution are employed to effect this), however, is more of a practical issue. In effect, the central and most basic theoretical question regarding PP and the requirement of precaution regards on what normative grounds we are to argue that a certain degree and price of precaution is the proper one to prescribe.

In Chapter 3, I argued that arguments for such normative validity need to move beyond claims about instrumental or formal rationality of the type typically made within traditional decision theory, and also include claims about *moral* validity. In Chapter 4, finally, I have tried to support the idea that this moral validity cannot plausibly be captured by traditional discussions and suggestions in normative ethics, but must also include claims about a special sort of moral quality attached to decisions to impose (or not impose) risks. The task of the present chapter is to develop and defend a theory about the nature and function of this quality.

---

[1] Simplistic conservatism tells us to preserve status quo for no better reason than it being status quo. Simplistic worship of the novel tells us to adopt/use any new thing, just because it is new.

C. Munthe, *The Price of Precaution and the Ethics of Risk*, The International Library of Ethics, Law and Technology 6, DOI 10.1007/978-94-007-1330-7_5,

In order to simplify the discussion, I will henceforth use the notion of the *responsibility of a risk imposition* (or *a decision to impose risks*) to signify the just mentioned moral quality. This use of the term 'responsibility' is different from some common uses within other philosophical discourses. In the classic discussion of free will and moral responsibility (or accountability), this term is used to refer to a quality of a person that needs to be in place if this person is to be capable of acting morally wrong (or deserving blame or similar responses on the basis of her action), regardless of the extent to which this action fulfils the conditions for moral wrongdoing according to an ethical theory of right action. The responsibility of a decision, in contrast, is first of all a property of this *decision* – i.e., neither of the person making it, nor of the factualistic outcome of the decision in terms of actual actions and consequences. Second, this use of 'responsibility' has less to do with basic theories of moral accountability or blameworthiness, than with the common sense idea that decisions and risk impositions may be morally appraised *in their own right*. The word 'responsibility' is quite commonly used to express opinions to this effect in common sense moral usage – as when we judge someone's behaviour to be reckless in spite of everything turning out well in the end. As we will soon see, the term has also been employed by philosophers that have tried to make theoretical sense of such common sense ideas.

I will start by sketching the basic structure of a theory of the moral responsibility of decisions to impose risks and relate this to a broader context of normative and/or moral inquiry. Based on this, I proceed to tackle two major challenges for any such theory: that of being sufficiently clear and practical to be able to guide action and that of being justified by good reasons. The former of these challenges is revisited several times throughout the chapter, while meeting the latter challenge is the main theme that gradually gives rise to a theory. I close by noting a number of potential problems with the sketched theory, nevertheless claiming these to be resolvable through further research.

5.1 Basic Structure

The basic notion of decisions to impose risks being more or less morally responsibly is, as indicated, not of my own making. As mentioned in preceding chapters, Rawls, in his argument for the maximin principle, alludes to what people would see as an irresponsible decision of the parties in the original position. Likewise, Jonas calls his own suggestion, above interpreted as a deontological rule, *the imperative of responsibility*. My suggestion is that this notion of responsibility should be seen as referring to a special and irreducible category of normatively relevant features.

Consider the following statements:

1. P acted (factualistically) morally wrong
2. P is a morally wicked person
3. The outcome of P's action was bad
4. P acted irrationally
5. P's decision to act as he did was irresponsible

All the types of facts described by these statements may be taken as reasons for some type of ultimate normative or moral justifiability – 'all things considered'.[2] Some of them may be weaker such reasons and some may be stronger, but I do not really need to specify this to any greater extent than what will transpire below. All that needs to be said at this initial stage is that statement 5 may be true although the statements 1–4 are all false, and that 5 may be false although 1–4 are all true. To explain this a bit further, consider how some of 1–4 might be true/false without the other ones being so. I may act irrationally but nevertheless morally permissible. I may act wrongly without being a wicked person. The outcome of my action may be bad without making my action irrational or morally wrong. And so on. In a similar manner, even if we do not know whether or not a person acted factualistically wrong or irrationally, whether he is wicked or virtuous or whether or not he produced a bad outcome, we may still know whether or not his decision to act as he did was a responsible one.

This last formulation in terms of what can be known indicates that, on a basic level, not only is the truth/falsity of 5 independent of that of 1–4; the ways in which we may inform ourselves about the truth/falsity of 5 does not necessarily involve informing ourselves about this regarding 1–4. This, in turn, is important as a clarifier of 5: it should not be conflated with the idea of acting or deciding from a set of motives, character traits or dispositions that are in turn evaluated positively from the point of view of any of the perspectives expressed by 1–4. In particular, to say that I acted wrong but that my decision to act in this way was nevertheless responsible is not to express the idea of 'blameless wrongdoing' (i.e. the idea that my wrongful act sprung from a set of disposition which I should have *according to the same principle judging my action as wrong*).[3]

Can this idea take us anywhere in our efforts to interpret the requirement of precaution and, in effect, to substantiate a more precise version of PP? In particular, can we do this in a way that will help us answer questions about how we should view cases like the one with the introduction of genetically modified crop described in Chapter 3? There seem to be two major factors that may counteract this. First, our base of intuitions regarding responsible risk-imposition may be too thin or confused to allow for the underpinning of any precise theory. Secondly, statements about the (lack of) responsibility of a decision to impose a certain risk may appear not to be

---

[2]Another possibility would be to view these statements as being about totally different subjects, lacking all connection to each other. This would not effect any change in what will be said below about what makes for a morally responsible risk imposition. However, due mainly to reasons having to do with the problem of guidance addressed below, I have come to prefer a model where all normative and evaluative statements are related to what they may function as reasons for.

[3]Cf. Parfit (1984, pp. 31ff). The reason, of course, is that this would again mean that the responsibility of risk imposition comes to be seen as a special case of the (factualist) rightness and wrongness of actions. More specifically, it would again actualise the two level approach – an idea that in the preceding chapter was rejected as a fruitful basis for justifying the requirement of precaution.

obviously action-guiding. In effect, if we believe that a certain action is morally right but that it would be irresponsible to decide to perform this action, it is doubtful whether this latter statement possesses any capacity to guide a decision maker. Let me begin by commenting briefly on this last issue.

5.2 The Problem of Guidance

If an action is (factualistically) morally right, it holds trivially that the performance of this action is morally justified regardless of whether or not the decision to perform it would be a morally responsible one. However, in most (if not all) situations where the requirement of precaution may apply, we will *not* be in the position of knowing which course of action is in fact the right one, the least harmful one, etc. In these cases, then, ideas about the responsibility of decisions in light of the risks they create may help to guide a decision maker. He may have no idea whatsoever regarding the rightness or wrongness of the resulting action, of course, but he might be able to substantiate the claim that, regardless of this, one decision would be more responsible than other decisions he could make. Or, at least, he may be able to rule out some options as being irresponsible to decide on.

What, then, if in retrospect we come to learn that, in fact, someone made the wrong choice? The idea of moral responsibility of risk impositions as a unique category of ethically relevant facts implies that if he did this through a responsible decision, he is less at fault morally or normatively speaking than if he had acted wrongly on the basis of an irresponsible decision. The same applies, of course, if his action turns out to be morally right. In this case, if his decision was irresponsible he is still at fault to some extent – although, obviously, less at fault than if his action had turned out to be wrong. In all, however, this type of retrospective judgements is not typically available to the decision maker at the time of deciding and can therefore not guide his decisions. Judgements about responsibility of decisions, however, may be so available and, therefore capable of guiding decision making.

This also makes clear how far an interpretation of the requirement of precaution based on a theory of responsible risk imposition can take us in the *justification* of decisions. Other things being equal, a responsible decision to impose some risk is always less faulty than an irresponsible such decision. However, there can still be valid complaints against this decision, its effects or the agent. For, as we have just seen, the action undertaken as a result of the decision may be (factualistically) wrong (although this was hard to know at the time of acting). And, even if that is not the case, we can still have virtue ethical complaints regarding the general character of the agent or simple complaints about the outcome of the decision being less than desirable. However, given the apparent facts that virtue ethics does not tell us much about which decisions to make, and that facts about the value of actual outcomes as well as the (factualistic) rightness and wrongness of actions are hidden from us in exactly those cases where the requirement of precaution seems to be most relevant, this complexity of types of moral complaints does not seem to tell against the idea of introducing into normative theory the further category of the responsibility of decisions to impose risks.

5.3 Basic Intuitions About Responsibility

Armed with these arguments, let us now turn to the first-mentioned challenge of trying to use intuitions about responsibility in the imposition of risks for the underpinning of a more specific theory. A successful attempt at this should provide us with enough ideas to argue in favour of a more precise interpretation of the requirement of precaution, which in turn might be able to help justifying the inclusion of some version of PP into, e.g., environmental policy or legislation.[4]

5.3.1 Absolutes or Degrees?

A first question to be addressed is this: Is the property of being responsible *absolute* or *gradual*? That is, can we describe a criterion in terms of one single property, which for any decision determines either that it was responsible or that it was not? Or is the responsibility of a decision to impose risks rather a property that is always present in various degrees in relation to the gradual occurrence of various other properties?[5] If we answer the first question affirmatively, it is possible to reformulate the forbidden risks approach discussed in the preceding chapter as a principle of responsible risk imposition rather than a deontological rule for the rightness and wrongness of action. For example, Jonas' imperative of responsibility may be reformulated in the following manner: If and only if a decision involves taking the risk of exterminating humanity some time in the future, this decision is irresponsible.

I believe, however, that our intuitions about responsibility – vague as they may be – tell strongly against such a conception. First, most of us would, I believe, want to make judgements about responsibility also when the risk of exterminating humanity is not an issue, such as in the case of my former neighbour the terrorist. And, I claim, the same would hold for whatever suggestion as to what outcomes it would always be absolutely irresponsible to risk.

Second, most of us would likewise want to assign different degrees of responsibility, for example, depending on what risks are imposed. This involves considering the degree of badness associated with the outcome-aspect of the risk as well the likelihood of the risk being actualised. For example, it makes a difference to how irresponsible my former neighbour was whether or not he stored his nitro-glycerine

[4]This further step, however, will need additional claims, for example, the political norm that society should involve itself in trying to make (certain types of) decisions more responsible. I will return to this aspect in the final chapter.

[5]This is really two different questions: (1) is the property of being responsible a binary property? (2) Is the occurrence of this property determined by one single property or relative to the occurrence of various other properties?

in movement-absorbing containers (making an explosion less likely). Likewise, it would also make a difference to the degree of responsibility of his decision if the risk he imposed involved a worse or less worse possible outcome – for example, if the risk involved destroying only my belongings, killing me, or killing me and ten additional people as well.

Third, I also believe that our intuitions strongly suggest that the degree of responsibility should be relativised to what *other* risks would be imposed by the available alternatives. For example, if the only alternative open to my former neighbour would have been to store the explosives in a way threatening not only me, but ten additional people as well, his actual decision would have been much less irresponsible (if not fully responsible). In other words, it might be quite responsible to impose many people to a very grave risk (in terms of harm risked as well as likelihood of actual occurrence of this harm), provided that alternative decisions would impose even graver risks.

Now, it may seem that by abolishing the absolutist approach, we lose out on a central element of normativity, namely the notion of an absolute category of irresponsibility that corresponds in the force of its implied moral disapproval to the notion of wrongful action. The gradual approach rather suggests a sliding scale of the type typically applied in theories of value, where there is no given point that demarcates valuable states from disvaluable states. This, in turn, might be seen as an additional threat to the prospect of receiving practical guidance from statements about the responsibility of risk impositions, since we cannot appeal to the idea of an absolute and given line between those risk impositions that are responsible and those that are not. Thus, we may seem to be impeded from ever making judgements of the kind 'it would be clearly irresponsible to decide to impose those risks!' And without access to such judgements, the prospect of any true guidance of practical action ever flowing out of a theory of morally responsible risk impositions would seem to crumble.

I agree that should this indeed be the consequence of the gradual model, this would be good reason to abolish either this model or the very idea of a theory of morally responsible risk impositions altogether. However, the gradual model still leaves room for *comparative* judgements of responsibility that are powerful enough to ground a notion of irresponsibility relative to a situation of choice, or – if there is need for comparing options from different situations of choice – relative to a particular set of possible decisions. Hence, we may say that, relative to such a set, a particular decision is irresponsible if and only if at least one other decision belonging to this set would be more responsible. And if we add to this the claim that an option is responsible if, and only if, it is not irresponsible, this secures that, for each situation of choice, there is at least one option that is the responsible decision in this situation. Compared to the absolutist approach, this model also has the advantage of allowing for the comparison of degrees of responsibility between irresponsible options, thus making possible the notion of one irresponsible decision being more or less irresponsible than another irresponsible decision.

Taken together, these ideas and intuitions seem to point in the direction of a theory living up to the following desiderata:

1. The responsibility of decisions is always a matter of degree.
2. In the choice or comparison between different possible decisions, those and only those decisions are irresponsible for which there is at least one other decision that would have been more responsible.
3. In the choice or comparison between different possible decisions that are all irresponsible, those decisions to which there is a greater number of alternative decisions that would be more responsible[6] are more irresponsible.[7]
4. There is no absolute quantity or quality of risk (such as the harm risked, the likelihood of this harm or a combination of these) such that any decision to impose such a risk is as responsible or irresponsible in all imaginable situations.
5. For any risk imposed by a decision, this decision can have varying degrees of responsibility (stretching from very responsible to very irresponsible), depending on the risk-magnitude in relation to the magnitudes of the risk impositions that would result from alternative decisions available in particular situations.[8]

[6]There may be problems involved in specifying this idea, due to the general problem of individuating and quantifying over alternative decisions, and the more specific possibility that the set of alternative decisions may be infinite. Handling these complications comes down to settling for one particular categorisation of an alternative decision capable of clearly separating different such decisions relative to a situation. Moreover, even if the total set of alternative decisions in any situation is in fact infinite, it does not follow that the subset of alternative decisions that would have been more responsible than a particular decision in the infinite (main) set must be an infinite set. Finally, in the present context it is not very important exactly how the idea at hand is specified. As will be seen below, what matters is that it may be specified in some way that makes the comparison of degrees of (ir)responsibility theoretically meaningful, so that, what might be called, 'quasi-absolutist' judgements about the (ir)responsibility of decisions may be inferred.

[7]This idea follows trivially from the gradual approach when we compare options in the same situation of choice. However, comparing options from different situations of choice, it may seem less obvious. In that case, it may be claimed, we should base the comparison on a thought experiment where we imagine the different options to belong to the same situation and compare the relative degrees of irresponsibility that would have been assigned to the options had this been the real situation of choice. This latter solution may seem to satisfy the intuition that the relative degree of irresponsibility should be determined solely by how options are ranked on the basis of those features determining the degree of responsibility (what these are will be further discussed below). However, my own intuition is rather that the interesting comparison to make regarding different irresponsible options is how irresponsible they are in the actual situation of choice. If they are equally irresponsible relative to their respective situations of choice, they are also equally irresponsible, period. The fact that they may have had a different relative degree of irresponsibility had they figured in *another* situation of choice is another matter entirely. Another idea is that it may make sense to have the comparison depend not only on the number of more responsible (or less irresponsible) options in the situation, but also on the number of less responsible (or more irresponsible) options. This would make the comparison of degrees of irresponsibility more complicated. I owe these observations to Ragnar Francén.

[8]This makes it possible to compare the degree of (ir)responsibility of a particular decision to impose risks, first, in relation to other particular situations of choice where this decision would or could be made, but also, second, between the decisions made in different situations of choice. One may ask why the latter is an important feature. I believe it is, since I believe that most of us have strong moral intuitions with regard to such comparisons. For example, compare the following two (clearly) irresponsible decisions made in different situations: (1) as a father I let my 3 year old

This excludes absolutist theories of the moral responsibility of decisions, while at the same time making room for 'quasi-absolutist' judgements of irresponsibility relative to given sets of possible decisions. However, from the idea that the degree of responsibility of a decision to impose risks depends on the relative magnitudes of those risk impositions we are to choose from we can also deduce that, in certain situations, such a decision may either lack all responsibility (and thus being maximally irresponsible) or be the only responsible one to make. More specifically:

6. If, in a situation, there is a decision such that it will reduce the magnitude of a risk without significant costs compared to alternative decisions the agent could make, it is always less responsible (or, in a particular context of this sort, irresponsible) not to make that decision.
7. If a decision to introduce a risk is to be responsible, this decision must either produce some sufficiently substantial benefit, or sufficiently reduce some risk.

That is, through points 6 and 7, the idea of degrees of responsibility provides good reason for criticising decisions to impose *unnecessary* as well as *unnecessarily grave* risks. This alone may be used for labelling as very irresponsible several decisions made for the sake of trivial pleasures or insignificant monetary gain, such as the introduction of new versions of consumer goods that introduce new risks but no new benefits to speak of compared to goods already available on the market. This point will be expanded on and seen to have some importance in the final chapter of this book.

Before going on to expand on these non-absolutist intuitions, however, a further possibility of invoking basic deontological-style ideas into a theory of morally responsible risk imposition has to be dealt with. This is the idea of according moral relevance to the intentions or motives from which an acting party acts, that was mentioned in Chapter 4 as the remaining possibility for an ethics of risks conforming to the standard deontological model to remain coherent.

5.3.2 What About Intentions?

Besides the idea of absolute bans, another central theme of deontological ethics is that of the moral relevance of *intentions*. According to this line of thought, there is a moral difference between the case where someone does something bad (say, causes someone to suffer) for the very reason of bringing this about and the case where someone does the same but not for this reason, although, perhaps, with a foreknowledge that the bad effect will follow. The former, of course, is seen as

child play around on the deck of a small sailboat at sea without any lifejacket on. (2) as the captain of a cruiser, I have neglected to see to it that there are lifejackets and lifeboats available for the 300 passengers aboard in case of an emergency in spite of the fact that I knew very well that the cruise will be very likely to pass through a severe tropical storm.

more wrong than the latter. Similarly, in order to *fully* live up to the requirements of morality, it is not sufficient to perform good or morally permissible actions, they must also be performed from the intention of doing this.[9]

To my knowledge, none of the more deontologically oriented commentators of PP (such as Hans Jonas) have tried to incorporate this idea into their suggestions regarding the morality of imposing risks. However, Per Sandin at times seems to suggest that incorporating a requirement of good intention would be a desirable feature of a plausible interpretation of PP.[10]

Sandin's reason for this position is that the concept of precautionary action, as applied in ordinary language, entails a requirement that such an action is performed from the intention of preventing whatever undesirable or harmful event relative to which the action is seen as precautionary in the first place. For example, if measures against the possibility of serious hazards in a chemical plant are to be called precautionary, the agent who applies these measures must act on the intention of preventing such hazards. On the basis of this, Sandin suggests that a sound interpretation of PP needs to take this linguistic fact into account, since PP will otherwise prescribe actions not properly seen as precautionary at all.

On my view, there are three partly interconnected reasons for rejecting this reasoning in the context of developing a theory of morally responsible risk impositions on the basis of which the requirement of precaution and, subsequently, PP may be clarified and defended.

First, there are strong reasons to doubt the generality of the linguistic practice cited by Sandin as evidence for his claim regarding the use of the word 'precaution' in ordinary language. True, in many cases of individual action, most people would indeed hesitate to call an action precautionary with respect to some undesirable outcome unless the element of intention was in place. However, if we look at those cases where PP is primarily applicable, these are not about the acts and decisions of individual people in singular circumstances, but rather about the extended activities of institutional agents, for example, the policies of a national authority or government. In such cases, I suggest, we do not normally pay much attention to the intentions behind these activities when deciding whether or not to call them precautionary. One reason for this is that it is problematic, to say the least, to ascribe determinate intentions to such agents. Of course, we may trace the decisions and considerations that have led up to the adoption of some institutional policy. But when doing so, we will end up with a rather complex picture where, due to political realities, the dominating intentions may in many cases be about securing electorate support, having a political opponent agreeing to some proposal in a completely different political

[9]This basic idea can be found in the writings of both Thomas Aquinas and Immanuel Kant, as well as numerous followers of the ideas of these philosophers. The idea is, of course, not entirely clear and can be interpreted in many different ways. One important distinction is that between requiring of the intention to be directed at that particular state or event that is condoned by morality (which seems to be Aquinas' idea), and requiring the intention to be directed at the abstract aim of doing one's moral duty (the core idea of Kantian ethics).

[10]Sandin (2004, pp. 11–15), Sandin (2005a) and Sandin (2007).

area, and similar things.[11] However, very few people would use *such* facts as reasons for doubting that the policy in question is truly precautionary. Consequently, such reasons would seem to be mostly irrelevant to the question of how precautionary are the activities prescribed by some version of PP, even if this question was to be approached solely as a matter of linguistic adequacy in relation to current usage.

Secondly and more important, even if I was to be wrong on the preceding point, this would have no bearing whatsoever on the question of what version of PP that *should* be adopted in actual policy making. Sandin does make the point that, in some instances at least, criticism of various versions of PP seems to be partly based on intuitions about whether or not they prescribe actions properly classified as precautionary.[12] Consequently, requiring PP to live up to such intuitions may seem to increase its plausibility. However, this claim at best seems to be valid regarding *other* features of the concept of precautionary action in ordinary language (as perceived by Sandin) than the requirement of intending the prevention of some hazard.[13] Thus, the most plausible understanding of this type of criticism is not that it feeds off linguistic intuitions about the concept of precautionary action, but rather that it is based on normative considerations that, as it happens, may also be shown to be incorporated into this concept. However, this is no reason for claiming that other elements of this concept, such as the requirement of intention, carry this type of normative significance.

As soon as this distinction (between evaluating PP on the basis of linguistic intuitions and evaluating PP on the basis of normative considerations that are also part of the concept of precaution) is clearly seen, it becomes obvious that the justification of PP and the requirement of precaution is entirely independent of investigations regarding the extent to which the actions prescribed by some version of these principles can be called precautionary or not. Perhaps they cannot, but it may nevertheless be the case that this is the version that we have most reason to accept – that prescribes the actions that should be prescribed according to a sound theory of morally responsible risk impositions.

This takes me to my third and final reason for rejecting Sandin's proposal. Simply put, I fail to find any reason for claiming a decision to impose some risk to be less responsible (and thus less defensible) just because this decision cannot be traced back to an intention of preventing some harm. Similarly, I fail to see why the fact that

[11]If the institutions are not political but, for example, commercial, other types of aims will transpire, such as the aim of maximising profit, keeping the shareholders happy, following the directions of a company board, et cetera.

[12]Sandin (2004, pp. 8–10), Sandin (2005a), and Sandin (2007).

[13]What Sandin points to is, first, that the concept of precautionary action entails a relativisation to some particular danger, hazard, undesirable event etc. This, in turn, can be used in order to rebut accusations against PP of implying absolutism or of necessarily being unable to deal with the balancing of different risks and chances. Secondly, he demonstrates how particular disagreements on precautionary policy (such as the differing views on the use of growth hormones in beef cattle between the E.U. and the U.S.) can be better understood with the help of the requirement that the agent possess "externally good reasons" for the various beliefs about the possibility of harm, et cetera, that underlies her action in order for it to be truly precautionary. See Sandin (2005a).

such an intention can be found among the grounds of some decision would constitute a reason for claiming this decision to be more responsible. Consider, for example, the case of a national government deciding on a policy that indeed does prevent possible harms but for the reason that adopting this policy will satisfy demands made by trade unions and that this, in turn, will make the latter claim less extensive wage increases, in turn seen as beneficial for the nation's economy. Assuming for the sake of argument that the degree of precaution expressed by this policy is acceptable in view of the price of precaution it implies, we may ask if there are any reasons for complaint against such a decision? Would it have been more desirable to have a different policy that would have prescribed a less than acceptable degree of precaution but from a clear intention of policy makers to prevent harm? Clearly not. Of course, as observed in the preceding chapter, we may complain against the policy makers that they are not acting out of a *virtue* of precaution. However, while perhaps constituting a ground for saying that they are less than nice people, this does not show that there is anything wrong with the decision they have made.

Having rejected also this deontological element, we are now in a position to consider further the implications of a gradualist approach to the moral responsibility of decisions to impose risks.

5.3.3 Assessing and Comparing Degrees of Responsibility

Points 1–5 in the gradualist approach set out above serve to highlight the fact that what degree of responsibility (or irresponsibility) is ascribed to a certain decision to impose risks must always be assessed through a comparison with other available risk impositions. However, so far we have not been provided with any hint regarding the basis for comparing different risk impositions with regard to the degree of responsibility of imposing or avoiding them. What factors are we to consider in order to assess such differences? In order to approach this issue, it is helpful to reflect on some imaginary cases of risk impositions, where some risks are created in order to alleviate other ones but in quite different manners and contexts. Let us, therefore, consider some cases of this sort:

> *Fred, the Desperate Commuter*
> Approaching the bus stop on his way home from work, Fred notices that his bus has just pulled up and started to take aboard passengers. He knows that if he misses this bus, he will have to wait 15 minutes for the next one and, once home, risk to miss the opportunity of playing for a while with his daughter before supper. Fred therefore starts running fiercely along the pavement of the heavily trafficked street, where many people, among them several children and elderly, are blocking his way. Cruising around them with wildly swinging arms, he is very close to knocking several of them off their feet, and some of these would, if so had happened, probably have fallen off the pavement and into the street, most likely in front of an approaching car. Luckily, although there is a close call, none of this actually happens and Fred gets on his bus, homeward bound at last.

Tony, the Saviour of the City
A very serious epidemic is plaguing a large city in a poor African country. Tony, who is the manager of a huge supply of antibiotics belonging to a pharmaceutical company of which he is the president, decides that he should help his fellow citizens. Using the resources of his company, he distributes the antibiotics around the city in spite of the fact that he knows that this action may contribute to a possible resistance of bacteria to this antibiotics some time in the future. Possibly, this may in turn contribute to the extinction of the human species some 10.000 years ahead.

Stan, the Lazy Builder
Leading the construction of a tunnel for trains and cars through a large ridge, Stan realises that the inner walls of the tunnel has to be lined with some material in order to avoid water leakage. There are two options – a comparably cheap chemical substance that can be swiftly sprayed onto the walls and the more expensive method of applying concrete. The latter method will also be more time-consuming, moving the time-schedule for the project six months ahead. The two methods are reported by their respective manufacturers to be effective enough. Stan, however, worries a bit about the chemical method, whether it might react with its surroundings and thereby let the water through as well as release poisonous agents into the ground and ground water. He asks the manufacturer about this, who denies any risk of this sort. He considers whether to consult some independent experts as well, but since it is Friday afternoon and since he wants to avoid any hassle from his bosses (who are eager for him to get on with the work), he decides to go for the chemical method on the basis that the evidence he has consulted shows it to be as safe as, somewhat more effective and cheaper than the concrete-method.

Granting the obvious fact that there may be relevant aspects of these cases not described above, my intuition is that Fred's decision to run was clearly irresponsible, while Tony's decision to distribute the antibiotics seems rather responsible. There are various factors contributing to this assessment. First, in the case of Fred, the risk he himself faces is not sufficiently serious to motivate the risks he imposes on the other people on the pavement. We may also point to the fact that he might have been able to run more carefully while still managing to catch his bus. However, even if we discount for that, Fred's decision was far from as responsible as it could have been. Tony, on the other hand, faces an acute situation with the threat of irreversible and very serious consequences (unlike Fred, who can play with his daughter tomorrow thus making up for the lost time today). He also knows his decision to be effective for meeting this threat. In view of that, the off-chance that his decision may some time in the future produce a more serious threat is no reason for judging his decision less responsible. Of course, if he had had the opportunity of choosing between distributing *two* types of antibiotics, one of which would help one person less while avoiding altogether the possibility of a future disaster, things would have been more complicated. That, however, is another story.

Now, Stan's decision is in a way more like Tony's than like Fred's. For, assuming the tunnel to be a desirable project,[14] the evidence seems to weigh heavily in favour of the decision he makes. However, in another way, Stan's situation is much different from both Fred's and Tony's, since he has time and opportunity to substantially improve his decisional basis of information without any risk of significant losses. This, in my view, makes his decision to carry on without making such an improvement rather irresponsible. In the light of possible future dangers resulting from our decisions, it is irresponsible not to investigate these, provided that this can be done without unacceptable costs.

What we learn from these cases is that three different factors seem to be involved in the comparison of different impositions of risks in order to assess their relative degrees of (ir)responsibility (apart from the consideration whether or not these impositions where unnecessary in the first place).[15] First, and perhaps most obvious, it is important to consider how undesirable it would be if the different risks where in fact actualised (i.e., the seriousness of the outcome aspect of the considered risks). This factor is what seems to tip the scale against Fred, since what makes his behaviour so irresponsible is the apparent lack of proportion between the possibility of missing an opportunity to play for a quarter of an hour with his daughter and the possibility of causing serious bodily harm to a number of people.

However, at the same time, this assessment obviously assumes the likelihood aspects of the risks faced by Fred to be reasonably on a par. That is, had the likelihood of Fred actually knocking anyone over in his rush to the bus stop been very slight (perhaps because there was only one person on the pavement), his decision to run in order to catch his bus may indeed have been quite responsible after all. This points to the importance of considering a second factor, namely the different likelihoods of the various risks being actualised (i.e., the magnitudes of the likelihood aspect of the considered risks). It is this aspect that seems to be the main reason for why Tony's decision can be assessed as quite responsible (and more responsible than Fred's, at that), since the threat he faces is one of more or less certain disaster while the risk he creates by alleviating this threat is very unlikely ever to be actualised. This tips the scale in Tony's favour even though the outcome aspect of the risk that he imposes is much more serious than that of the risk that his action serves to eradicate.

Third, we come to the importance of considering not only the relative magnitudes of the risks in question, but also the quality of the decisional basis underlying the assessment of these magnitudes and – even more importantly – the possibility and

[14]If it is not, the decision to build the tunnel is irresponsible for the simple reason that it unnecessarily introduces new risks.

[15]Taken together, these three factors imply that my theory of the morality of imposing risks, and the version of the requirement of precaution, as well as any version of PP based on this theory, is not implied by any of the suggested generic versions of PP that Martin Peterson has claimed to be incoherent (Peterson 2006). In addition, my theory does not satisfy the 'dominance' or 'covariance' conditions assumed by Peterson in his proof of the just mentioned incoherence. All of this depends on the simple fact that, on moral grounds, I reject the idea that only outcome values and likelihoods are relevant from the point of view of a sound ethics of risks.

costs of improving this quality. This is what we see in the case of Stan; his decision is made irresponsible by the very fact that he could quite easily have improved the quality of his decisional basis without any significant cost or production of additional risks of harm. Furthermore, the force of this factor is illustrated by the apparent fact that had Stan's assessment of the situation been made on a better decisional basis, or had the costs of improving his actual decisional basis been more substantial (in terms of actual loss or additional risks), his decision could very well have been seen as quite responsible.

Armed with these – admittedly very unclear and tentative – suggestions, let us now take a renewed look at the case described in Chapter 3, with the genetically modified crop designed to avoid serious environmental damage, but also bringing unclear risks for future ecological disaster.

This case is a bit like all of the cases above. Like in the case of Fred, we have the opportunity of avoiding something harmful. Like in the case of Tony, the harm that can be avoided is irreparable and significant, and the risk facing us involves substantial harm but a very uncertain likelihood. Like in the case of Stan, we have the opportunity of improving our decisional basis regarding the risks we face. However, although such an improvement will be associated with some costs also in Stan's case, in the present case these costs are far more substantial than some loss of money and time. On the other hand, the costs are not as immediate as they would be in the case of Tony.

On the basis of this and the intuitions collected so far, it seems clear to me that it would be irresponsible to introduce the new crop on a broad basis immediately, without any investigation regarding the likelihood of ecological disaster. However, the apparent fact mentioned when I introduced this case above, that the responsibility of deciding on further investigation has a limit, seems to remain. If we choose the responsible path of improving our decisional basis, and if this investigation corroborates the hypothesis that a catastrophe can be avoided, somewhere along the line it will become irresponsible not to start using the crop (in the way which makes the catastrophe scenario less likely).

From this, we can try to look at similar cases, where various parameters are modified in a way that moves us closer to present realities. Suppose, for example, that the genetically modified crop has no potential for reducing current environmental destruction in farming, but only to make the use of certain pesticides more monetary cost-effective (since the crop has been made resistant to this particular pesticide). This will, of course, change the scene dramatically, since lengthy investigations about the possibility of future ecological disaster will not mean that seriously harmful processes are allowed to continue. In terms of the aspects of the requirement of precaution influencing the degree of precaution it prescribes, this can mean either that we employ stricter proof-standards than in the former case or that we widen the set of scenarios counting as examples of "too serious risks", for example by taking more remote possibilities into account in the "may bring great harm" clause. The fact that these factors can vary from case to case is a direct consequence of the idea described above of making the responsibility of imposing risks relative to the particular circumstances of the situation under consideration.

I will return to the application of my theory of morally responsible risk impositions to practical cases in the final chapter. Before that, however, we need to consider two more questions: First, is it possible to mould the rather vague skeleton of a theory outlined above into something more sharp and precise? In particular, could we find reasons to support a fully naturalised criterion of the moral responsibility of decisions to impose risks, stating clear conditions in terms of quantifiable natural properties that determine all possible instances of degrees of responsibility and comparative relations of responsibility between different risk impositions? Second, are there any further factors beside those three mentioned above that should be introduced as influencing the degree of responsibility of risk impositions? As we will see, these questions are interconnected in important ways.

5.3.4 Avoiding Indeterminacy – Possibility and Desirability

In the first chapter, I commented briefly on the idea that the primary virtue of any interpretation of PP is a high degree of exactness and precision. In particular, I referred to the idea that a satisfactory version of PP must be able to produce (given access to relevant facts) clear-cut answers as to what alternative is recommended, from the point of view of precaution, for any situation of choice.[16] If this requirement is to be met by the approach to the clarification of PP developed in this book, the same degree of exactness and precision must be found in my suggested theory of the morality of imposing risks. This due to the general structure of my approach, that a satisfying version of PP must express a sound version of the requirement of precaution, which in turn must prescribe proper degrees and prices of precaution, which in turn presupposes that its prescribed decisions are cases of morally responsible risk impositions.

However, the account given above of those factors influencing the moral responsibility of a decision to impose risks and how this influence works is far from being close to the exactness just described. On the contrary, it would seem that the theory permits quite large portions of indeterminacy as to the relative degree of responsibility of various decisions to impose risks. That is, for many pairs (or larger sets) of possible or actual decisions to impose risks, the theory will be unable to rank unequivocally which risk imposition is the more or less responsible one. In consequence, we will have a significant number of cases where there is no answer to questions of the type 'would this risk imposition be more or less (or as) responsible than (as) that one?' All we can say is that it is true of each of them that they are neither more nor less (nor as) responsible than (as) each other.

There are several sources of this indeterminacy. One is that the factors taken to influence the degree of responsibility are vague. This is partly a consequence of my willingness to allow into the theory qualitative (or, at least, non-numerical) assessments of likelihoods and values. More important, however, seems to me the

[16] Sandin and Hansson (2002). This idea is also implied by the 'total order' condition set out in Peterson (2006).

inclusion of the third factor of the quality of available evidence (i.e., the idea that failure to improve one's decisional basis when this can be done at reasonable costs makes for a lesser degree of responsibility of one's decision to impose some risk). Of course, statistical theory can here be brought in to specify in very precise numerical terms aspects such as confidence intervals. However, this cannot help us to bypass the kind of qualitative assessments of the relevance of various pieces of evidence that have to be undertaken when the available evidence, as is so often the case, does not consist of earlier instances of *exactly the same* type of events as the one we are trying to predict.[17] And, finally, even if these assessments could be moulded into numerical exactness in a satisfactory way, this would still leave the issue of where the thresholds of different levels of *quality* are located (i.e., where a leap has been made so as to make the body of evidence seriously *better*). In addition, it remains to specify how considerations of such quality are to balanced against the costs and risks of improving it and, furthermore, how the results of such balancing are to be related to the general picture of likelihoods and values at stake in order to produce a final assessment regarding the moral responsibility of risk impositions in a particular situation of choice.

This kind of indeterminacy in a theory may be seen as problematic for different reasons. One such reason has to do with practical applicability – in the present case, the problem of extracting actual guidance regarding what decisions should be prescribed by PP. I will return to that problem in the final chapter and there argue that this problem is actually much less severe than what is often assumed. Another reason, however, has to do with the scientific or theoretical virtue of preciseness. It may be claimed that if a theory is indeterminate or imprecise, it cannot be seen as one particular theory at all, but rather as a family of several mutually incompatible theories. That is, deciding on such an alleged theory will not be a case of adopting any specific and coherent view at all, but merely to point more or less vaguely in some more or less unspecified theoretical 'direction'.

In my opinion, depending on the field of study, this may in many cases be quite sufficient – at least when judged as an isolated accomplishment marking the start of a further chain of development towards greater clarity.[18] More importantly, however, even if we were to lack all reason to believe such a chain to be forthcoming, indeterminacy would still be quite compatible with a fully satisfactory theory. For, on reflection, even though most of us hope for reality to contain fine-grained structures

---

[17]This is the realistic picture to paint regarding available evidence when considering possible risks and benefits of previously unapplied technologies – especially regarding applications on larger scales where significant effects on complex systems, such as the ecosphere, may be anticipated. In these cases, we will often have quite a lot of evidence about various seemingly relevant mechanisms taken in isolation (or in more limited combinations), but lack completely any experience of the workings of all relevant mechanisms taken together in exactly that kind of surrounding in which the application is being pondered (partly because we will have to take seriously the possibility that some such mechanism is unknown to us until we acquire experience of this application).

[18]It should be noted that this increased clarity may consist in the insight that the idea of a fully deterministic theory of the moral responsibility of risk impositions is a fundamentally hopeless prospect.

possible to describe in very precise and deterministic terms, this may turn out to be a hope in vain. At least, some *degree* of indeterminacy may turn out to be an essential constituent of some *parts* of reality – such as the parts where heaps of sand and bald people reside.[19] Consequently, should our normative and evaluative intuitions about the morality of imposing risks suggest such vagueness to reside in this particular area of 'normative reality', not to let this be reflected in our theory would be a cardinal mistake. Just as, in the case of empirical theories, empirical adequacy trumps all other kinds of scientific considerations, when we talk about moral theories, actually making the morally adequate evaluations and prescriptions is the supreme aim. All other theoretical aspirations, such as simplicity or, as is the target of my present argument, conformity to basic hypotheses of determinism, play second fiddle whenever the basic normative soundness of an ethical theory is at stake.

On a theoretical level, therefore, indeterminacy is something that, in many cases, we should be able live with quite easily. Indeed, should it turn out to be a necessary price for attaining normative or moral acceptability of a theory of the morality of imposing risks, we should receive it with a warm welcome.

None of this is to suggest, of course, that exactness and precision are to be seen as worthless features. On the contrary, if evaluative and normative adequacy can be retained, higher degrees of exactness and precision is a desirable objective. However, in order to be able to assess if such an objective is a realistic aim in the framework of my approach, we must first consider whether there are any further morally relevant considerations that could serve to make the theory of morally responsible risk impositions more precise. This since the chief source of indeterminacy identified above seems to be the indeterminacy of exactly those factors influencing the degree of such responsibility. In the next section, therefore, I will proceed to investigate the possibility of making the theory more precise through the consideration of additional features influencing this.

5.4 Areas of Precaution

What, then, about the more exact balancing of various risks and possible benefits? Can anything more precise be said about this from the point of view of intuitions about responsible risk imposition, thus providing a basis for a more precise version of the requirement of precaution? As has been demonstrated earlier, this question can be rephrased as a question regarding the proper degree of precaution required in different kinds of situations, this degree in turn being influenced by a multitude of factors. Moreover, the desiderata of avoiding decisional paralysis and arbitrary

[19]For those not familiar with philosophical discussions of vagueness, characteristics and phenomena such as these constitute standard examples of very familiar and obviously true aspects of reality that are vague (i.e. we are unable to say exactly what number of hair strands must be lost in order for the feature of baldness to appear, or what number of grains of sand need to be assembled in order for a heap to have been formed).

recommendations can be said to pinpoint extreme forms and degrees of precaution that should not be prescribed by the requirement. Taking this as illustrating the extreme end of a sliding scale, we can also describe its opposite end as complete lack of precaution – i.e. complete or exaggerated discounting of the fact that some activity may bring great harm, which in turn makes for a very irresponsible decision. In between these extremes, we find a large area of various degrees of precaution, in relation to which we can ask where a morally responsible decision to impose risks should be located.

Moreover, I believe that the intuitions demonstrated above also point to a further restriction, namely that a harm of a certain magnitude should always be seen as *at least* as important as a benefit of the same magnitude. That is, apart from considerations having to do with likelihoods and qualitative aspects of the harm and benefit (such as, for example, the harm or benefit being of an especially serious kind, or considerations of justice in the distribution of the harm or benefit), it is at least as irresponsible to impose a risk of harm of magnitude x as it is to prevent the occurrence of a chance of a benefit of the corresponding magnitude. In effect, such a prevention should be seen at least as an equivalent (from the point of view of responsibility) imposition of risk of harm. This idea can, I believe, be further expanded into the idea (including also considerations of likelihood) that the creation of a risk with a certain *expected* (negative) value should be seen, other things being equal, as *at least* equivalent to the prevention of a possible benefit with a corresponding expected positive value (since this makes the negative expected value of this prevention equal to that of creating the risk). This, in turn, enables us to add to our sketched scale a middle point of what Sven-Ove Hansson has called *risk neutrality*,[20] where equal expected values are seen as equally important from the point of view of responsibility. We can then locate the available area for possible responsible degrees of precaution in between such risk neutrality and extreme degrees of precaution[21]:

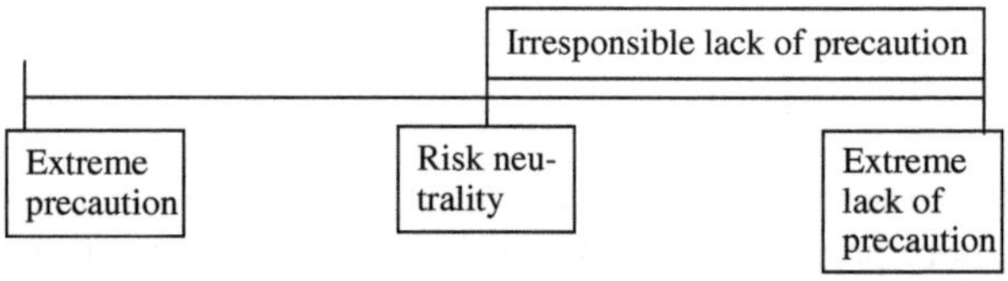

As pointed out by Hansson,[22] it may be that the vague idea about more of precaution being desirable in human decision making (that is the 'precautionary ideal' mentioned at the outset of this book) boils down to nothing more than a complaint about *actual* decision making having been residing in the area of irresponsible lack of precaution (e.g., due to so-called technology optimism) and should move left up

[20]Hansson (1999).

[21]Cf. Munthe (1997, chapter 5), and Hansson (1999).

[22]Hansson (1999).

to the point of risk neutrality. This is a claim that is in full compliance with the standard idea of rational action as calculated risk taking – i.e., the idea that within decision theory is expressed by the principle of maximising expected utility. This due to the fact that, according to this principle, we should ascribe equal normative weight to harms and benefits with equal products of magnitude and likelihood – i.e., given equal likelihoods, a harm of a certain negative magnitude (such as –1) and a benefit with the same positive magnitude (such as +1) balance each other out perfectly (so as to produce the sum of zero) and nothing more influences what decision we should make. In consequence, if we have to choose between one option bringing a risk of harm of –1 with probability 0.5 and a chance of benefit of +1 with probability 0.5 and another option bringing a risk of harm of –10,000 with probability 0.5 and a chance of benefit of +10,000 with probability 0.5, these options are completely on a par in terms of responsibility from the point of view of risk neutrality.

However, the model also allows us to say that, perhaps, responsible precaution should move beyond risk neutrality and closer to extreme precaution (although, of course, never all the way). This raises the question what could speak in favour of such a move.

5.4.1 Beyond Risk Neutrality

A good reason for attempting a theory about the morality of imposing risks that places the conditions for a responsible decision beyond risk neutrality is that the idea of risk neutrality can be demonstrated to be flawed from an ethical point of view. This is the claim that I will now make and defend.

Actually, the example just used to illustrate what the idea of risk neutrality in this context boils down to supplies us with the skeleton of an argument to this effect. What is shown by this rather formal explanation is that risk neutrality fails to consider what affected parties stand to win or lose by choosing different risk distributions with equal expected values. This complaint can be fleshed out further to demonstrate its moral leverage. Consider the following case:

> *The Risk Neutral Provider*
> You are the provider of a family of five children and an unemployed partner. They depend entirely on your pay-check to have access to food, housing, education, et cetera. Having collected this week's salary, you are now on the way home when you are being intercepted by a street artist offering you to play a lottery where the price of a ticket equals your pay-check (say, €100), but where you are given a 50% chance of doubling it. Alternatively, you may disregard the offer, proceeding home to your family with €100 to pay for the coming week's needs. The expected outcome of entering the lottery is thus (0.5 × 0 + 0.5 × €200 = €100), while the expected outcome of proceeding home is (1 × €100 + 0 × €0 = €100). Thus, if you are assumed to be risk neutral, you would consider the choice between these two options to be indifferent. According to risk neutrality, whatever you do, your decision will be equally morally responsible. But, clearly, this is not the case! As a matter of fact, entering the lottery would be a blatantly irresponsible decision under the circumstances.[23]

[23] As with all examples of this sort, we may of course complicate it; assuming uncertainty as to the quality of the lottery, further values than merely money, et cetera. However, as everyone clever

It is important to be clear about what is illustrated by this case. The point is not to rule out the broad idea of calculated risk taking. We could change the outcome- and likelihood-numbers so that the moral conclusion would be less clear on the irresponsibility side with regard to entering the lottery, but where the reasoning giving this conclusion would still be a case of calculated risk taking. Rather, the point is that the idea of risk neutrality completely fails to take into account the obviously relevant consideration that the family provider has certain moral responsibilities with regard to avoiding the possible negative outcome for his or her family.

Connecting this judgement to the general characterisation of the notion of risk neutrality made earlier, we can now see that the flaw just illustrated is general. Simply put, a plausible idea of the morality of imposing risks needs to avoid the idea of necessarily having decisions with equal expected values being judged as equally morally responsible. Furthermore, the moral responsibility of decisions to impose risks has to be determined to some extent by the outcome values connected to the expected values of risks, but independently of the latter. Such a theory moves beyond risk neutrality in its moral assessment of risk impositions.

This raises the question of how such a theory may be achieved. In order to justify a move beyond risk neutrality, it needs to be demonstrated that this move can be engineered in a satisfactory way. There seems to me to be two main types of blueprint for such a venture. One of these will be presented and rejected in the next subsection, while the remaining one will be the subject of the following main section.

5.4.2 The Quality of Available Evidence

It may be tempting to use the idea of the irresponsibility of not improving one's decisional basis when this can be done at acceptable cost to motivate that risks with different outcome values but equal expected values should be treated differently. At least, this may seem an attractive suggestion regarding options where disaster-like outcomes (i.e., very large negative values) are involved in the risks created. This since even a small adjustment of the likelihood of a disaster will make for a significant change of the expected value of the corresponding risk. In effect, the assessment of expected values in such a case can be claimed to be extremely *unstable* in the sense that it is very likely to change for the worse in the light of further evidence. Consequently, if such an adjustment can be made at reasonable cost, it may be claimed that it would be irresponsible to opt for an option that brings such a risk of disaster on the basis that the expected value of this option is equal to that of an alternative option that does not bring such a risk.

However, on further reflection, I find it hard to defend such a suggestion as a general claim unless we can say that risking disaster-like outcomes is something

---

enough to point that out realises, every such move may be counteracted by further changes (the safety of the way home, even more preferences pulling in the opposite direction than the formerly added ones, and so on).

that is extra undesirable in its own right.[24] This is due to the combination of two facts. First, in examples such as these, we have to assume that the risk of disaster is balanced by either a chance of considerable benefits or of very low likelihoods (otherwise, the option bringing a risk of such a disaster could hardly have the same expected value as an option not bringing such a risk). Second, I still stand by the claim made above that a harm of a certain magnitude should always be seen as *at least as important* as a benefit of the same magnitude. Therefore, we also have to add to the calculus the factor of missing the chance of large benefits and, if we do that, the just set out argument would seem to loose much of its force. For just as small adjustments of the likelihood of large disasters make for large changes of expected value, the same goes for similar adjustments regarding the likelihood of large scale benefits.[25]

On the basis of this, it seems to me that appeals to our reasons for improving the quality of available evidence could at best have rather limited implications for the kind of cases presently considered. In particular, it provides no reason to apply a higher degree of precaution than what is implied by risk neutrality.

Admittedly, the argument set out so far may support unequal treatment of options with equal expected values in cases where such equality of two options, only one of which brings a risk of a disaster-like outcome, is due to a very low likelihood of this outcome actually occurring. In such cases, the magnitude of the risk is still very likely to increase in the light of further evidence without the presence of a corresponding likelihood of increasing a chance of extreme benefits. However, to this we must also add the possibility that the additional evidence would actually make for a *decrease* of the risk of disaster. That is, the idea of us having a strong responsibility to base decisions regarding risk impositions on better evidence (when this can be done at reasonable cost) does not really tell us anything about how risks of different types of outcomes should be balanced against each other. Just as an improvement of the quality of a body of evidence may make for a decrease of the expected value of some option it may also effect a change for the better.

True, we may still argue that *what is to count as acceptable costs* for improving our decisional basis can be heavily affected by the presence of disaster-like outcomes in the risk-picture of some option. If we run the risk of creating a disaster, this may seem as a very good reason to allow for larger costs of collecting further evidence in order to double-check that the disaster-risk of this option is indeed sufficiently low due to very low likelihood of actual occurrence or balanced by a sufficiently large chance of sufficiently large benefits in order for it to be recommendable all things considered. However, again, this would seem to be no reason to abandon the position of risk neutrality. Rather, it merely underlines the importance of treating the action of improving our decisional basis as just another option among those we are choosing among and that, consequently, the risks and chances of this

[24]This seems to be the vague idea of Allhoff (2009). Below, I argue in favour of a more refined, flexible and less binary/rigid version of such a general intuition.

[25]I owe this point to Toni Rønnow-Rasmussen.

option have to be balanced against the risks and chances of the other options open to us in order for our decision to express a proper degree of precaution. To be sure, such a calculus must also take into consideration the importance of basing decisions on information of sufficient quality. But this in itself tells us nothing about whether this balancing should be made according to the position of risk neutrality, or if it should express a greater degree of precaution.

In conclusion, in order to acquire reasons to move beyond risk neutrality towards a greater degree of precaution, we seem to be forced to consider the possibility of some deeper ethical anomaly being hidden within the basic idea of calculated risk taking (expressed by the decision theoretical principle of maximising expected value). In particular, we need to investigate the possibility of questioning the basic tenet of this idea that harms and benefits of equal corresponding magnitudes are always equally important from a moral point of view.

5.5 The Weight of Evil

This brings us to the idea that it is more important from a moral point of view to alleviate or prevent some particular types of harm than to create some particular types of benefits.[26] That is, even if two options have equal expected values, the nature of the components from which these expected values are aggregated may matter when it comes to the assessment of the *moral importance* or *weight* of these expected values. More precisely, if one of these expected values is the result of an aggregation of possible outcomes among which there is a particular type of harm (and the other one is not), this expected value is assigned an extra negative weight from a moral point of view. In effect, it is more important to avoid the risks created by this option than those of the other one. As a consequence, it is irresponsible to choose this option in spite of the fact that its expected value equals that of its alternative.

Here is a simple example reminding why such an idea is worth considering as an alternative to the ideal of risk neutrality. Remember the implication of this neutrality that the decision to impose the two risks (0.5×1) and (0.5×-1) is as morally responsible/irresponsible as a decision to impose the risks (0.5×10.000) and (0.5×-10.000). This numerical and abstract illustration does not tell us anything very informative from an ethical point of view, however, so let us flesh out this general implication in a case with more concrete and obviously morally sensitive features. Assume that you, for some unfortunate reason, are in the situation that you have to make a choice between two lotteries, L1 and L2. L1 implies receiving a ticket for free with a 0.5 likelihood of having to work as a slave for 1 week and an

[26]This idea presupposes that we define the concepts of harm and benefit, so that they may be distinguished in theory. Since there are ways to define these respective notions so that this assumption cannot be made, I thus presuppose the concepts of harm and benefit to be other ones than these. This aspect is further elaborated on below.

equal likelihood of winning enough money to be economically independent for a week. L2 implies receiving a ticket for free that gives you a 0.5 likelihood of having to work as a slave for the rest of your life, but also an equal likelihood of being economically independent for the rest of your life. Most of us, I presume, would not only opt for L1, we would also judge anyone's decision to choose L2 as particularly ill-considered.[27] Now, let us change the case only slightly, so that now you are making a decision affecting not your own life, but someone else's (who we assume to possess no additional morally relevant features). With this change, I take it, the apparently sound judgement of a decision to choose L2 as being ill-considered is qualitatively sharpened into a clearly moral complaint of *lack of morally appropriate consideration* with regard to the affected party. But this judgement cannot be explained by the idea of risk neutrality, since the expected values of L1 and L2 are assumed to be equal. What may explain this judgement, however, is the idea of an increased moral weight of evil: L2 is the inferior choice due to the presence of a risk involving an exceptional type and quantity of harm.

5.5.1 Conceptual Preliminaries

A necessary basic assumption for making sense of this idea is that it is possible to distinguish the concepts of harm and benefit. That is, it has to be assumed that there is at least one way of defining these notions such that not every harm is also a benefit and not every benefit is also a harm. There are many ways to accomplish this. One obvious strategy, hinted at in Chapter 2, is to base the demarcation between these concepts on a theory of value. For instance, employing a hedonistic account of the good and the bad, we may say that harm occurs whenever some suffering is introduced into the world or some pleasure is being prevented from being so introduced. Conversely, a benefit occurs whenever some pleasure is introduced or some suffering is prevented. A similar scheme may be employed for other accounts of the good and the bad.

Another strategy is to explain the difference in terms of changes on some value-variable in comparison to an index. For example, instead of formulating a hedonistic theory of value based on a conceptual distinction between pleasure and suffering, we may want to express it in terms of a continuous variable of subjective experiential well-being. This variable may have a zero-point (such that if my life as a whole is below this point, it is not worth living, or that if a moment of my life places me on this very point, that moment makes my life neither better nor worse) that can serve as an index. So, if an experience is introduced into my life that either pushes the

[27] Unless, of course, that person can convince us of a preference or value ordering where the difference between being economically independent and living in slavery is not a significant one. As this is not what people think outside of philosophical thought experiments, this possibility is disregarded in the following. It should further be observed that variations regarding how economical independence and slavery is compared in terms of (un)desirability is irrelevant, since whatever comparison seems apt has to be done in the assessment of both L1 and L2.

value of my life closer to the zero-point on the positive end of the scale, or pushes it farther away from this point on the negative end of the scale, this experience is a harm. Had it instead functioned in the opposite way, this experience would have been a benefit. This idea, in turn, may relate itself to other indices, such as the value of my life before the introduction of the experience in question, or this value compared to a possible world where the experience does not enter my life at all. Such further indices may also facilitate a distinction between the concepts of harm and benefit even in the absence of a zero-point. Again, this formal scheme may be adapted to other theories of value than those belonging to the hedonistic family.

Obviously, all of these ways of distinguishing the concepts of harm and benefit are compatible with, e.g., a concept of *harming* that uses both of the other concepts (the same goes for the concepts of benefit and benefiting). For instance, we may want to say that I am *harming* someone also in a case where my action is that I withhold a benefit from someone. This only goes to show the importance of keeping apart the concept of harm and the concept of harming (indicated by the fact that the latter concept is indicated by a verb, while the former one is not), and similarly for the concepts of benefit and benefiting. For if we did not, we would be unable to express the just mentioned idea.

More importantly, the just made point highlights the fact that distinguishing the concepts of harm and benefit is also compatible with any normative view on the moral weight of evil. At the same time, in the present context, the job of such a distinction is to facilitate a normatively plausible idea about this weight that can motivate a move beyond risk neutrality. This illuminates a response to any suggestion that the concepts of harm and benefit should not be distinguished (that is, we should define them so that they are extensionally equivalent): From a methodological point of view, if distinguishing these concepts can be demonstrated to be a part of a plausible moral theory, such suggestions are unwarranted.

In conclusion, therefore, it is the arguments for and against the plausibility of a particular moral idea on the weight of evil, and a resulting moral ideal of precaution that extends beyond risk neutrality that will decide what concepts are fruitful to employ. The rest of this chapter will be about different versions of such ideas and various arguments that can be wielded for and against them.

5.5.2 Five Approaches

There are a number of ways in which an idea regarding an increased 'weight of evil' may be construed in the context of an ethics of risk. As a matter of fact, we have already encountered one of these – namely the forbidden risks approach that was discussed both in the foregoing chapter and earlier in the present chapter. According to this view, certain evils (i.e., types of adverse outcomes) have an infinite negative moral weight, so that the presence of such an outcome in the risk-picture of some option makes it absolutely irresponsible to choose that option – no matter what other harms or benefits may be alleviated or achieved.

Since I have already found good reason to reject this particular suggestion, I will not consider it further here. However, this leaves at least five further ways in which we might construct an idea of an increased negative moral weight of some types of evils or harms. We may pursue:

1. *Strong rigidity*: assign one fixed (though not infinite) extra negative weight to one type of harms.
2. *Weak rigidity*: assign different extra negative (though not infinite) weights within a set of different selected types of harm so that each such weight is fixed for each type of harm.
3. *Rigidity of aggregation (Strong or Weak)*: assign (fixed or variable) extra negative weights to certain types of harm when these are to be balanced against risks of certain other types of harm (including lost chances of benefits).
4. *Simple progressiveness*: assign different extra negative (though not infinite) weights to harms independent of their type but proportionally to their magnitudes.
5. *Relative progressiveness*: assign different extra negative (though not infinite) weights to harms relative to what other harms and benefits are at stake in the particular case and independent of the types of harm and benefit but proportionally to their magnitudes.

Note that for each of these approaches it holds that the extra negative weight may never be infinitely large (since that would reintroduce the forbidden risks approach). That is, each of the approaches concedes that, no matter the type or magnitude of harm, it is always theoretically possible for it to be counterbalanced by the possibility of alleviating some other type of harm or the chance of achieving some benefit of sufficient magnitude or importance. However, in addition to this, I will also assume that the extra negative weights assignable within the different approaches may never be *extremely large*. This in light of the criticism put forward of the forbidden risks approach, that also a weakened version of this idea (where imposition of the forbidden risks can only be justified if there are extremely large benefits or harm preventions in sight) is open to the charge of prescribing a much too high price of precaution.

Admittedly, the five approaches are rather different. Moreover, each of them may be varied within a rather large spectrum as to the magnitude of the extra negative weights they assign and the conditions for the assignation of these weights. For example, the strong rigidity approach may be specified in terms of a variation of Jonas' "imperative of responsibility" (discussed in the preceding chapter) – assigning a rather substantial extra negative weight to the outcome of the eradication of the human species. But it may as well be specified either as assigning a much smaller weight to this type of outcome, or as assigning the extra negative weight (whatever its magnitude) to some broader type of harm (including that implied by the eradication of the human species), or some other type of harm altogether. However, in my view, the different approaches can still be sorted into two different main groups sharing significant features.

The first three 'rigidity approaches' all share the feature of assigning extra negative weight to harms on the basis of some *qualitative* grouping of different harms into different *types*, quite independently of the *magnitudes* of these harms (this magnitude being determined by the applied theory of value). For example, in the just outlined variation of the strong rigidity approach inspired by Jonas' imperative of responsibility, outcomes of the type demarcated by the description 'the human species is eradicated' are assigned a very large extra negative weight just because they fall under this description and in complete disregard of the fact that harms of other types may be much greater. In the case of the other two rigidity approaches, the design may be much more complex – since these approaches permit a much higher degree of nuance and variation in the assignation of different extra negative weights to different types of harms. However, they still build on the basic idea that the extra moral weight is to be assigned on the basis of qualitative differences rather than differences of magnitude. In this way, the three 'rigidity approaches' may all be seen as retaining an essential feature of deontological ethical theories,[28] although, in this case, in the context of the morality of imposing risks rather than the (factualistic) rightness and wrongness of action.

In contrast to this group of approaches, the two 'progressiveness approaches' base their assignation of extra negative weights to harms exclusively on the magnitudes of these harms. As explained in the beginning of Chapter 2, these magnitudes may vary quite significantly depending on what theory of value is applied. However, given the application of one such theory, the progressiveness approaches work exclusively from the magnitudes of different harms thereby implied. No further division of harms into different types on the grounds of qualitative description or classification is needed. In this, both of the progressiveness approaches may be seen as retaining an essential feature of standard consequentialist ethical theories.[29]

I will now argue that the progressiveness approaches are superior to the rigidity approaches. That is, the assignation of extra negative moral weights to harms on something else than the magnitudes of these harms constitutes a serious shortcoming. After that, a special subsection will be devoted to the rigidity of aggregation approach and the connected idea of basing an ethics of risk on the notion of basic moral rights. Having rejected also this idea, I will then go on to defend the claim that the relative progressiveness approach is superior to the simple progressiveness approach.

5.5.3 The Case Against Rigidity

My general objection to the rigidity approaches is really quite simple: they all fail to take adequate account of what makes harms morally important, namely their

[28] See the beginning of Chapter 4 for an explanation of what is traditionally taken to unite such theories.

[29] Again, see the beginning of Chapter 4.

harmfulness – the fact that they make things *bad*. No matter what theory of value is applied (and thus what idea of what is to count as harms and what determines the magnitude of harms), the morally important thing about someone being harmed is that this individual is made worse off. Moreover, the more worse off someone is made – i.e., the more harmful the harm she is subjected to – the more morally important it is to have this harm prevented, alleviated or at least reduced. In consequence, when comparing harms in terms of moral importance, the relevant thing is their relative magnitudes: the greater the harm, the worse the moral fault of inflicting it. For example, if I inflict a certain amount of suffering on someone (and we assume this to be harmful), this is bad since it harms this person and thereby makes her worse off. This is in itself a reason for me to abstain from inflicting this suffering. However, had I inflicted an even greater suffering, or inflicted the same suffering for a longer time, this would have been even worse. Thus, in such a case, the moral reason for me not to inflict this harm would have been even greater. Reversely, had I inflicted a lesser suffering or as intense a suffering for a shorter time, this would have been less bad – thereby making for a weaker reason for me to abstain from this harm-infliction (although, of course, I still have such a reason).

These, I claim, are the basic features of harms that make them morally significant: other things being equal, harms make the world a worse place and greater harms make it even worse. But this is flatly denied by the rigidity approaches. According to all these ideas, one harm may be worse than another harm although the harmfulness of the former is lesser than that of the latter. Thus, these approaches may all imply prescriptions to inflict greater harms in order to avoid smaller ones. And this simply because the latter fall under this or that qualitative description.

Philosophically educated readers may sense a parallel between this argument and that of those hedonistic utilitarians who object to British nineteenth century philosopher John Stuart Mill's so-called qualitative hedonism. Hence, the debate between those who side with British eighteenth century philosopher Jeremy Bentham's famous claim regarding any activity that...

> ...the value which they possess, is exactly in proportion to the pleasure they yield. Every other species of pre-eminence which may be attempted to be established among them is altogether fanciful. Prejudice apart, the game of push-pin is of equal value with the arts and sciences of music and poetry. If the game of push-pin furnish more pleasure, it is more valuable than either.[30]

And those subscribing to Mill's equally famous statement that...

> It is better to be a human dissatisfied than a pig satisfied; better to be Socrates dissatisfied than a fool satisfied.[31]

However, although there is a parallel, the argument just outlined really resides on a higher theoretical level than the dispute between 'Benthamists' and 'Milleans'. What these quarrel about is what theory of value is to be accepted – that is, what is

[30]Bentham (1907). Quotation taken from Singer (1994, p. 200).

[31]Mill (1993). Quotation taken from Singer (1994, p. 203).

to count as harms and benefits and what determines the magnitudes of these. What Milleans claim, and Benthamists deny, is that certain pleasures may be of greater benefit than others although they are equally pleasurable and that, correspondingly, certain displeasures may be more harmful than others although they are equally unpleasant. My argument, however, does not presuppose any particular position on this matter. What I claim is that given any definite theory of value – i.e., account of what determines the magnitudes of harms – what determines the *moral importance* of harms is their magnitudes (i.e., how bad they are) and nothing else than that.

This leaves the possibility open of putting forward a *theory of value* that introduces some form of rigidity in the very determination of harmfulness. Consequently, I do not deny the possibility of defending the view that the relation between degrees of occurrence of those natural features that are taken to make the word better or worse and the corresponding degrees of goodness and badness is not a linear one. To take a simple example within the framework of hedonism, one unit of suffering (whatever that is) may be taken to make for one unit of badness while two units of suffering make for *three* units of badness. Or, to state a simple version of Mill's qualitative hedonism, one unit of one type of suffering may be less bad than one unit of another type of suffering.

However, *no such theory would challenge the position of risk neutrality*, since that idea merely states that corresponding magnitudes of *badness and goodness* balance each other out in terms of moral importance, that two combined events of equal magnitudes of *badness* make for double the magnitude of negative moral weight, that the same holds regarding *goodness* and positive moral weight, etc. Thus, introducing rigidity in the step between factual description and the evaluation of facts in terms of goodness/badness would not advance the idea of an increased moral importance of evil.

5.5.4 Rigidity of Aggregation and the Notion of Rights

This dismissal of all rigidity approaches right across the board may appear overly hasty, however. For one particular version of these approaches may in fact be taken to imply an idea that actually *can* be combined with the view that the moral importance of harms depends exclusively on their harmfulness. This is a version of the rigidity of aggregation approach that is a traditionally central feature of all ideas of basic moral rights: In order to have moral leverage, comparisons of harmfulness need to observe properly, as Rawls once put it, "the distinction between persons".[32] Hence, the degree of harmfulness that makes for a morally more important harm has to be a harmfulness *for some particular person*. For this reason, not any trade-off of possible harms can be seen as capturing something of moral importance. In particular, *interpersonal aggregates* of individual harms cannot be the basis of morally

[32]Rawls (1971, p. 27).

relevant trade-offs against a harm to an individual. This basic idea is what separates rights-based and consequentialist approaches to trade-offs between competing individual interests.

In recent decades, such a "person affecting view"[33] has been demonstrated to actualise quite peculiar problems when applied to the issue of ethically assessing the effects of our actions on future people and generations,[34] as well as the ethics of human reproduction.[35] Citing Derek Parfit's famous example, since a large scale reform to the benefit of the future environment may be expected to influence heavily who will exist in the future, the person affecting view may prescribe us to prefer short term overconsumption of natural resources rather than a policy that would leave generations and generations of future people with the chance of having access to these same resources.[36] This "non-identity problem" transports itself directly into the very core of rights based ethics,[37] suggesting that the ethical notion of rights is ill-equipped as a moral basis for our dealings with the future – in particular regarding the sort of large-scale issues with a high impact on posterity to which PP and the requirement of precaution are meant to apply.

This general reason to distrust the idea of the person affecting view being applied within a rigidity of aggregation approach to the morality of imposing risks may be further complemented in the present context. In relation to the discourse that has given rise to the idea of a PP and the notion of an ideal of precaution, the issues of levels and time-horizons, briefly mentioned in Chapter 2, add further arguments for resisting the notion of rights as being part of the basis of a sound theory of the morality of risk impositions. To see this, we need to briefly review the core of the basic notion of moral rights.

As mentioned, one part of this core is the person affecting view – the idea that the moral seriousness of harm to a person can only be balanced by an equally or more harmful harm to some other person. Translated into a language of risk, this equals the idea that the imposition of a risk to a person can be morally responsible only if it is necessary to avoid an equally (or more) morally serious risk to another person.[38] This idea allows for applying the logic of rights-based ethics to the ethics of risk without being forced to adopt the forbidden risk approach. Thus, we have a room for that idea about *prima facie* or "defeasible" individual rights against risk impositions that Sven-Ove Hansson has recently claimed to be central for a sound ethics of risk.[39]

---

[33]Parfit (1984, p. 370).

[34]Parfit (1984, chapter 16).

[35]See, e.g., Munthe (1999c, especially chapter 5).

[36]Parfit (1984, p. 362).

[37]Parfit (1984, pp. 364–366).

[38]If this condition is dropped, what results is a typical consequentialist view regarding morally relevant trade-offs between the impositions of risks to different parties.

[39]Hansson (2009).

Another part of the conceptual core of the notion of a moral right is captured by the notion of *waiving*. Any holder of a right has the power to cancel its moral importance in a situation, if she so pleases – rights, in the words of Ronald Dworkin, are "trumps" to be wielded *at the discretion of the right-holder* in order to have actions of others become morally problematic. This in contrast to such duties of others towards a person that are not based on her having a right; these may very well continue to be of moral importance even if the person in question honestly declares that she does not mind having them violated. Hence, the very *concept* of moral rights contains a minimal autonomy component: rights need only be respected to the extent that right-holders want to have them respected.

Both of these components have problematic implications in the context of PP and the requirement of precaution due to the level and time-horizon dimensions of precaution mentioned in Chapter 2. As observed, both these dimensions imply the possibility of many risk impositions, that each taken by itself is morally defensible (or meets the conditions of the requirement of precaution) add up to a total mass of risk impositions that is not defensible. For instance, when pouring the last rest of chlorine after a house cleaning session into the sink, I may not impose much of a risk to anything of importance, while a regulation banning such behaviour brings some salient extra complications into my life, thus imposing on me a salient (though not very serious) risk. Consequently, should I be recognized as having a right not to have risks imposed on me, such a regulation would violate this right without the presence of a comparable risk to any individual thereby being prevented (assuming other people to be in a relevantly similar situation to me). Of course, the regulation will have very important *overall* effects in risk reduction terms through its influence on the behaviour of a mass of people. However, those effects are aggregated outcomes of all these individuals acting together. For each of these individuals, it still holds that the regulation imposes a risk on them that is not balanced by a resulting risk reduction of comparable importance to anyone else. To say that the right is balanced by the overall risk reduction resulting from the regulation, we would have to abandon the person affecting view, thus not recognizing me as the holder of a basic moral right against risk impositions.

Suppose further that the point just made is accepted, and that the person affecting view is therefore abandoned as a basis for the ethical assessment of risk impositions. Thus, there is good reason for implementing the regulation in terms of its overall risk reductive effects. A supporter of the rights-based perspective of an ethics of rights may then suggest that we may at least talk about a *semi-right* being protected. Every individual has a (semi-)right to having this overall risk reduction being achieved, since they will then be able to share this collective good. Conversely, we may say that withholding the ability to share such a good from them would violate a (semi-)right to being exposed to a reasonable overall risk-level. However, in order for it to make sense to talk about this reason in favour of the regulation even in terms of a semi-right, we would need to hold on to the component of waiving. Otherwise, what we have is just a consequentialist reason of the classic sort: it is responsible to implement the regulation due to its overall effects compared to the downsides in terms of personal comfort for affected individuals.

To see how the component of waiving brings problems in this context, we need only assume that each of the individuals waives this semi-right. What is then left is the (fully fledged) right of individuals not to have their personal comfort disturbed. *No other valid reason for or against the regulation is at hand*, in spite of the fact that failing to implement it may severely affect basic societal functions, to the detriment of all. That is, waiving is similar to the person affecting view in that it provides individual people with an unconditional veto that may be wielded against obviously powerful moral reasons.

To conclude, then, the general rigidity approaches fail since they are unable to account for the basic idea that equally harmful harms are equally morally important. The rigidity of aggregation approach may account for that idea in a way that suggests a (non absolute) ethics of rights as a fitting form for the morality of risk impositions. However, this approach fails due to its inability to respond adequately to the fact that precaution and responsible risk imposition in many cases is a collective rather than individual good. All of these factors, however, can be accounted for by the progressiveness approaches, since they do not attempt to distinguish the moral importance of different harms or benefits on the basis of anything else than their badness or goodness, and since they contain neither the person affecting view nor the notion of waiving. Let me, therefore, now turn to these approaches in order to further investigate the possibility of defending the idea that morally responsible risk impositions may require a higher degree of precaution than mere risk neutrality.

5.5.5 Simple Progressiveness

According to the simple progressiveness approach, harms become steadily more morally important as they become more harmful, however, at an increasingly faster pace, so to speak. That is, a curve in a system of coordinates describing the raise of moral importance (measured on the Y-axis) in relation to increased degrees of harmfulness or badness (measured on the X-axis) would have to have a steeper angle than 45°, or, if the curve is not a straight one, would have to take the form of an exponential function with a steeper upward progression than that of a 45° straight line. All this assuming, of course, that the relation between *goodness* and moral importance would, in such a figure, be described by a 45° straight line. That is, the relationship between units of goodness and moral importance would be one of 1:1, while relationship between units of badness and moral importance would be one of $1{:}1{+}n$.

As an illustrating example, we may consider the idea of assigning to each extra unit of harmfulness a fixed extra negative weight of moral importance in the form of factor 2 (so that for one unit of harmfulness we get two units of moral importance, for two units of harmfulness we get four units of moral importance, for three we get six, etc.). This idea can then be demonstrated by the following figure:

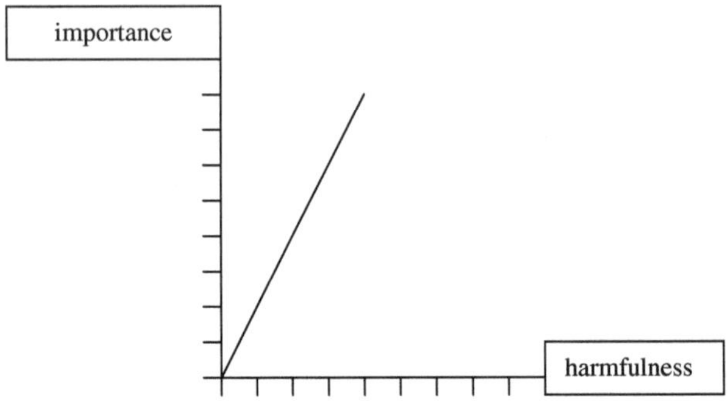

Proceeding on the assumption that the relationship between goodness and moral importance is one of 1:1, this means that it would take double the amount of goodness in order to balance the moral importance of any amount of badness. That is to say, in the choice between two options, one of which effects a certain amount of harm and the other one does not, the first of these options would also have to effect at least double the amount of additional goodness in order for this option to be morally justified. Complicating the picture a bit further, if option o_1 would effect x units of badness and option o_2 would effect x units of goodness, o_1 would also have to effect $3x$ units of goodness in order for these options to be morally on a par (two of which are taken to balance out the one unit of harm and the third to balance the one unit of goodness of the other option) and even more additional goodness if o_1 is to be morally preferable to o_2. This in contrast to the case where we assign no extra negative moral weight to harmfulness, in which case $2x$ units of goodness effected by o_1 would be sufficient to make this option morally on a par to o_2.

Relating this idea to the discussion of the morality of imposing risks, let us now consider what difference an idea like this would make in contrast to the position of risk neutrality. Suppose that we have to make a choice between two options, each of which having two possible outcomes that would occur with the following likelihoods (designated by the letter p) and involving the following amounts of goodness and badness (designated by the letter v):

$\boldsymbol{o_1}$
Outcome 1 : $p = 0.2/v = +2, -80$
Outcome 2 : $p = 0.8/v = +120, -5$

$\boldsymbol{o_2}$
Outcome 1 : $p = 0.8/v = +100, -15$
Outcome 2 : $p = 0.2/v = +50, -20$

On the risk neutrality approach, all we have to do in order to rank these option in terms of moral responsibility is to calculate their expected values. For o_1 this amounts to 76.4, while for o_2 it equals 74. Thus, on this approach, o_1 is the more morally responsible decision.

On the idea of assigning to harms the extra negative weight of factor 2, however, the calculus becomes more complicated. For we now have to consider the fact that the various possible outcomes involve different degrees of harmfulness, the moral importance of which are double that of the various possible benefits. In order to account for this in the calculus, we may simply multiply with factor 2 the value of each possible negative upshot while the beneficial upshots are left standing as they are. This gives us the following matrix:

o_1
Outcome 1 : $p = 0.2/v = +2, -160$
Outcome 2 : $p = 0.8/v = +120, -10$

o_2
Outcome 1 : $p = 0.8/v = +100, -30$
Outcome 2 : $p = 0.2/v = +50, -40$

And if we now repeat our calculation, we see that the adjusted negative values make for an adjusted expected value for o_1 of 56.4, while that of o_2 equals 58. That is, this calculus makes for a reversed judgement regarding what decision would be the more responsible one.

It should be emphasised that these *adjusted* expected values are in fact *not* the expected values of these options – those are the ones first calculated above. Rather, what has just been demonstrated is that accepting the idea of an extra negative moral weight of evil in line with the simple progressiveness approach may effect that it is morally responsible to choose an option with a lower expected value than alternative options. That is, while in the just outlined example the risk neutrality approach would judge it responsible to risk a rather large harm (+2, –80) that will occur with a rather low likelihood (0.2) in order to achieve with a rather high likelihood (0.8) an even greater benefit (+120, –5), on the version of the simple progressiveness approach presently under consideration, this would be irresponsible. In effect, this latter approach here prescribes a higher price of precaution since it tells us to cash in a 0.8 chance of gaining a large benefit for the same chance of gaining a much lesser one and a 0.2 chance of gaining even less in order to avoid the 0.2 likelihood of substantial loss.

In order to put some actual flesh on these rather abstract bones, we may imagine that we are managing a petro-chemical company and face the choice of either immediately embarking on a new method of production that bears with it a promise of producing a less environmentally detrimental form of gasoline at a somewhat lower cost (o_1), or to postpone this introduction in order to further investigate and, if necessary, safeguard against some risks of unintentional emissions of highly poisonous gases due to possible anomalies in the new technology (o_2). The actualisation of this risk is represented by outcome no. 1 of o_1 in the tables above, while a successful immediate introduction is represented by outcome no. 2 of this option. Outcome no. 1 of o_2 instead represents the situation where the introduction is successfully made at a later date (therefore bringing a lesser gain and introducing additional costs for the investigation), while outcome no. 2 instead represents the possibility of such actions becoming necessary – thereby further increasing costs and reducing benefits. What we have seen is that the position of risk neutrality prescribes us to embark on the new production method immediately, while our presently considered version of the simple progressiveness approach instead urges us to proceed with more caution and run the risk of much lesser benefits and increased costs in order to avoid a not very likely disaster-like outcome.

To be sure, this example is an extreme simplification of the kind of circumstances that occur in real life. Equally true, it is also rigged in order to demonstrate the points just made. That is, had the numbers been just slightly different, the upshot may very well have been that the two approaches would have concurred in their implications. However, the example still suffices to illustrate how the simple progressiveness approach differs from the position of risk neutrality. Moreover, as I will now try to show, it also serves as enough of a basis for pointing out some important weaknesses of simple progressiveness.

5.5.6 The Case for Relative Progressiveness

A basic fault of the version of simple progressiveness outlined above may be claimed to be its insensitivity to the apparent fact that the extra moral importance of harms should, if it is to make any sense, be more closely connected to the harmfulness of harms. On the idea of adding to each harm an extra factor 2 of additional moral importance, the smallest of harms is assigned the same extra negative moral weight as the most horrible disaster we may imagine. Moreover, this feature will remain whatever simple number we choose for such a fixed factor.

This apparent flaw can, it may seem, be avoided by increasing more radically the progressiveness of the extra negative weight. We may, for example, assign such a weight according to the formula x^x, or $x(2+x)$, thereby ensuring that the additional moral importance increases progressively as the harmfulness of the harm increases. However, increasing the rate of progressiveness of the moral importance of harm along such lines may at the same time be seen as a weakness, since it makes the idea of an extra negative moral weight of evil more susceptible to the objection that it distorts the importance of what *benefits* are to be found in choosing *alternative* courses of action. This serves to highlight what, in my view, is the real problem with the simple progressiveness approach.

In my view, to be ethically defensible, the moral importance of running the risk of harm or loss has to be seen as relative to not only the harmfulness of that particular harm but also to *what we have to lose or gain by choosing any of the other options open to us in a situation of choice*. For example, if we are really bad off, running the risk of becoming even more worse off in order to be able to reach a decent quality of life is not as morally serious as if we already enjoyed this level of well-being and tried to reach an even higher quality of life. Hence, if I in my present condition of relative prosperity was to decide on a course of action in order to make myself even better off and thereby run the risk of making myself destitute, this risk would be much more morally serious than if I had tried for the same (or even a slightly lesser) improvement of well-being, running the same risk of becoming destitute, from a position of a life just barely worth living. This is the basic intuition that is captured by the idea of relative progressiveness: the moral importance of harms becomes greater the more we have to lose by embarking on an activity that may lead to the harm in question.

This underlying idea shares some similarities with John Rawls' notion of "the priority of liberty".[40] According to this view, people who are interested in their own well-being and rational will be prepared to sacrifice some personal freedom in order to advance their own well-being, but only to a certain point. For example, they may accept restrictions of political liberty in order to advance a course of socio-economic development that promises a higher rate of growth in terms of available well-being (the fruits of which they are able to reap). At some level of well-being, however, this preparedness vanishes and they become unwilling to cash in any amount of liberty for additional gains of well-being. That is, the negative moral weight of losses of liberty varies in relation to what these people have to gain or lose in other respects.[41]

In my view, Rawls' priority of liberty model is problematic due to its use of an absolute threshold beyond which the negative moral weight of liberty restrictions becomes infinitely great. That is, in this model, when a certain level of well-being is secured, even the smallest restriction of liberty becomes impossible to justify no matter the size of the gain in terms of well-being. In contrast, the relative progressiveness approach works within the limits set out above, that the additional negative moral weight of harms can never become infinite or even extremely large. Moreover, Rawls' idea is shaped on the basis of a separation between the values of liberty (or freedom) and well-being that is not employed by the relative progressiveness approach.[42] On this latter idea, the only acknowledged separation between values is that between benefits and harms (i.e., goodness and badness).

Apart from this, however, the ideas share the basic feature of relativising the assignation of additional negative moral weight to harms to what is at stake through the other options open to the choosing party. If these mean that she may escape an intolerable situation by risking an even worse fate, taking this latter risk is less morally serious than if she had a decent level of well-being secured (either as a

[40]Rawls (1971, §82).

[41]Rawls himself claims that the ultimate rationale of this reasoning is based on liberty or autonomy as the ultimate value, since the reason why the choosing parties will be willing to sacrifice political liberty under the conditions mentioned is that these conditions impede an effective use of the freedoms guaranteed by such liberty. However, what this actually means is that liberty or autonomy are viewed as instrumental (and not ultimate) values by the parties whose choice justifies the priority of liberty construction.

[42]In this, the priority of liberty idea is more close to the 'rigidity approaches' rejected above, since it bases its assignation of negative moral weight to evils on a differentiation of different types of harms (loss of liberty as opposed to loss of material welfare).

matter of fact or as a consequence of some of the other options she faces) and took the risk of disaster in order to advance even further her quality of life.

In order to illustrate how this idea may work in connection to a risky decision of closer relevance to the typical practical context of PP, we may consider the following imaginary (and still heavily simplified) case: As in the example presented earlier in this book about the genetically modified crop, we assume that our current farming practices of using fertilisers and pesticides are seriously detrimental to the environment. If this practice is allowed to continue, very serious harm will ensue. However, just stopping the practice will also cause serious harm, although of a different kind (due to recession of the farming economy, extremely elevated prices on farming products, serious shortage of such products, etc.). As before, we further assume that bioscientists present us with newly developed genetically modified crops that may produce a sufficiently rich harvest at a sufficiently low cost in order to avoid this latter consequence. However, again as before, the large scale use of these crops also bring unclear long-term risks of ecological disasters much more serious than the threatening detrimental consequences of our current practices. Now, let me supplement these assumptions with the following one: the scientists also present us with a program of research and control that can serve to investigate more precisely the nature and magnitude of the risk of ecological disaster and, when this is done, to safeguard against such devastating outcomes so as to make them extremely unlikely.[43] At the same time, this program will, of course, make use of substantial resources and take quite some time to implement. The decision makers in charge thus seem to face four options (below which are listed the chief risks and chances effected by them):

o_1: *Go on as before*

1. Retained productivity and production costs
2. Almost certain very serious harm due to environmental degradation

o_2: *Stop the use of fertilisers and pesticides*

1. Serious harm due to loss of productivity
2. No harm due to environmental degradation

o_3: *Immediate large-scale introduction of genetically modified crops*

1. Immediate avoidance of very serious harm due to environmental degradation
2. Retained productivity and production costs

[43]The application of these safeguards, we assume, presupposes detailed knowledge about the various mechanisms involved in the creation of the risk of ecological disaster.

3. Unclear risk of extremely serious harm due to long-term ecological disaster

o_4: *Delay large-scale introduction of genetically modified crops and implement program of research and control*

1. Retained productivity and production costs
2. Retained harm due to environmental degradation during the implementation period
3. Avoidance of very serious harm due to environmental degradation after the implementation period
4. Extra costs due to research program
5. Possible extra costs due to the implementation of safeguards
6. Very small likelihood of extremely serious harm due to long-term ecological disaster

The presently considered idea is that the size of the extra negative moral weight assigned to bad outcomes should depend not only on the magnitudes of each of these outcomes *taken by themselves*, but also how these magnitudes compare to the magnitudes of the possible outcomes of the other options in a situation of choice. If one of these options provides *a mix of risks and chances that makes for a decent or acceptable outcome even if worst was to come to worst*, possible outcomes of the other options that are even worse are assigned a significantly larger extra negative weight than if this option had not been open to the deciding party.

Let us now assume, for the sake of the argument, that o_4 actually provides such an acceptable or decent mix of risks and chances. To be true, this option introduces some extra costs (some of which are merely possible) and forces us to retain the environmentally degrading farming practice for some time. However, after this time, it provides us with a practice that avoids environmental degradation and retains farming productivity while, at the same time, making sure that the likelihood of extremely bad side-effects is significantly reduced. On this assumption, it seems clear that the extra negative weight assignable to the unclear risk of o_3 must be much greater than if o_4 had not been open to the deciding party. It is, of course, an open question whether or not this increase of the extra negative weight is large enough to make o_4 clearly more morally responsible than o_3. This will, among other things, depend on the length of the time-period necessary for the implementation of the research and control program, the level of environmental degradation that will ensue as a result of this and the amount of resources that needs to be spent on the program. However, according to the presently considered idea of relative progressiveness, the chances of o_3 being a morally responsible decision becomes much less due to the presence of o_4 among the alternatives open to the deciding party.

It is a part of this idea that if we lose the alternative of effecting an acceptable mix of risks and chances, we also lose the reason for increasing significantly the extra negative moral weights of the harms risked through the remaining options. The natural idea would then seem to be to employ either the risk neutrality approach or a moderate version of simple progressiveness (the latter solution will make for a higher degree of precaution also in the absence of alternatives that would secure an acceptable mix of risks and chances). It is, however, not very clear to me whether this choice would make any significant difference in my example. It still would be highly unclear what decision among o_1, o_2 and o_3 that would be the more responsible one. However, if o_4 is again assumed to be open, this would effect not only an increased extra negative weight of the harmful possible outcome of o_3, but of the harmful possible outcomes of o_1 and o_2 as well (although, of course, not equally drastic increases). And, in the light of this, o_4 would be more responsible than both o_1 and o_2. In order for it not to be, the harmful aspects of this option would have to be so harmful that it would become doubtful if it should really be considered to secure an acceptable mix of risks and chances. Now, we have already seen that o_3 need not be a responsible choice even in the absence of o_4. Moreover, the increase of the negative moral weight assigned to the possible harmful upshot of this option in light of the presence of o_4 is even greater than that assigned to the harmful possibilities produced by the other two options. Taken together, these alleged facts would seem to strongly suggest that, as long as o_4 would deliver an acceptable mix of risks and chances and is open to the deciding party, it will also be more responsible than any of the other options.

5.6 Problems with Relative Progressiveness

There are a number of aspects of the relative progressiveness approach, as outlined above, that need to be clarified. Each of these may also be seen as highlighting potential problems.

5.6.1 What Implications for other Normative Issues?

If we accept the relative progressiveness approach in connection to the issue of what determines morally responsible risk impositions, does this mean that we have to accept corresponding ideas also in connection to other normative areas of inquiry? That is, will my suggested answer to the issue of the moral responsibility of decisions to impose risks have implications for, e.g., issues about what is to be seen as virtue, what determines the rightness and wrongness of action, et cetera?

In my view, the answer to this question is negative. As explained at the outset of this chapter, it is a part of the basic structure of the theory I am trying to defend that its truth or validity is independent of the truth or validity of theories addressing other issues than that of the moral responsibility of risk impositions. I also stated that the prospect of justification of such theories (i.e. what reasons there are against them or in their favour) is independent of the same prospect regarding the theory that I have set out above.

In spite of this, the following may seem an appealing line of thought. Suppose that we are able to make a case for a theory like the relative progressiveness approach in the context of formulating factualistic criteria for morally right- and wrongful actions.[44] Would not this strengthen the case in favour of the relative progressiveness approach also in the context of the moral responsibility of imposing risks? After all, this would give us access to the claim that both these versions of the relative progressiveness approach are supported by a more general idea regarding the moral weight of evil. And if this idea could be shown to support versions of the relative progressiveness approach also in the context of virtue ethics and, maybe, even in the context of the theory of rational action, this would seem to strengthen the case in favour of the relative progressiveness approach in the context of the moral responsibility of risk impositions even further.

I remain cautiously open to this suggestion, appealing as it is due to its promise of greater coherence among our normative thoughts in different areas. However, the case made above in favour of the relative progressiveness approach is not dependent on any such further argument. It rests solely on basic intuitions regarding the moral responsibility of imposing risks and I am not at all sure whether or not I would subscribe to similar convictions in other areas of normative inquiry. In my view, this is a virtue of my suggestion, since it serves to retain the assumed independence of the issue of the moral responsibility of risk impositions from other moral and normative issues.

5.6.2 The Lack of Numerical Exactness

Despite the occasional use of exact numbers in the descriptions of risks and chances involved in some of the scenarios considered in this chapter, I am not presently prepared to formulate the relative progressiveness approach with such a degree of precision. One simple reason for this is that I have failed to find any grounds for choosing one particular numerically exact version of this approach rather than another. This does not imply that I am prepared to accept any version of the relative progressiveness approach. As stated above, I am not prepared to accept such versions where the extra negative moral weight assigned to harms is infinitely or extremely large. Moreover, I would be prepared to argue that it must also not be so small that the difference between the relative progressiveness approach and the position of risk neutrality becomes insignificant in the vast majority of practical cases.

This means that the relative progressiveness approach fails to annul all of the indeterminacy in the theory of morally responsible risk impositions that was noted earlier in this chapter. In spite of the fact that this approach clarifies that the proper degree of precaution indeed exceeds risk neutrality (at least in some circumstances),

---

[44] For an interesting attempt in this direction with applications in the area of energy production, see Arrhenius and Bykvist (1995).

it remains unclear exactly how much it does so. As argued above, I do not see this as a devastating objection on a theoretical level. On the contrary, the apparent lack of reasons for settling on one particular numerically exact version of the relative progressiveness approach seems to fuel the suspicion that some facts regarding morally responsible risk impositions are indeed vague. That is, in some circumstances there is no clear answer to the question whether or not a particular imposition of risks would be responsible or not.

However, besides this, I actually believe that the lack of numerical exactness is in fact a virtue. This since it makes room for comparisons of different options in those situations where likelihoods and values cannot be numerically represented in a reasonable way. This connects to what I see as a cardinal weakness of most models developed within decision theory – their ambition to present numerically exact criteria of rational actions and decisions (and thereby requiring information on this level of exactness) makes them inapplicable to most practical situations. In the case of my theory, however, it is possible to assess all situations of choice regardless of what kind of information is available. To be sure, there remains the possibility of indeterminacy in some cases as to what is actually recommended. However, in the next chapter, I will argue that such indeterminacy is compatible with the good enough practical guidance of decision making being provided.

5.6.3 What Size of the Weight?

A key issue for any idea to the effect that evil upshots should be assigned extra negative moral weight is, of course, how large this weight should be. This problem is actualised no matter if we go for some of the rigidity approaches, simple progressiveness or relative progressiveness. It may be argued that this unclarity is, in fact, devastating for all such ideas. For, even on reflection, it may seem extremely hard to provide convincing reasons for adopting one size of the weight rather than another. And, in the light of this, any version of the basic idea of an extra negative moral weight of evil may be accused of being arbitrary and thus suffering a lack of rational foundation.

The problem with this argument, however, is that it seems to have equal force with respect to the position of risk neutrality. Appealing as the idea of assigning equal moral weight to equal magnitudes of goodness and badness may appear due to its theoretical simplicity and symmetry, it is still the case that it is one position among many on a sliding scale of different possible moral weights assignable to harms and benefits. Another position on this scale is to assign extra *positive* moral weight to (certain kinds of) *benefits* – or, which amounts to the same thing, *less* negative moral weight to harms as compared to the positive moral weight of equally large benefits. Thus, what we are dealing with seems to be a continuum of possible assignations of moral weight to harms and benefits and, on this continuum, the position of risk neutrality is as arbitrary as any other suggestion in the absence of good reason.

For sure, many ethicists, philosophers and risk researchers would protest that the idea of an increased moral weight of goodness (or a decreased such weight of badness) is clearly inferior to the position of risk neutrality. They would argue that basic moral convictions help us to see this alleged fact. However, such a line of reasoning may equally well be repeated in support of a particular version of the idea of assigning extra negative moral weight to evil. That is, the dispute regarding the size of this weight must, ultimately, be settled by a moral argument that appeals to basic moral convictions. This is exactly what I have done at the end of the preceding section, the outset of the present one and when I have argued that the weight must be neither infinitely or extremely large, nor so small that it becomes insignificant in practice. The fact that I personally at this point have been unable to come up with any further basic moral convictions that might have warranted a more precise choice regarding the size of the weight is no reason to claim that such an accomplishment is impossible. On the contrary, in my view, it is a challenge for further research in the area of the ethics of risk.

5.6.4 Pure or Mixed Relative Progressiveness?

This unclarity has already been briefly touched on above and arises exclusively within the context of relative progressiveness. According to this suggestion, if one of the options open to the choosing party would secure an acceptable or decent mix of risks and chances, the possible harmful upshots of the other options are assigned extra negative moral weights. But what if no option would actually secure such a mix? Should we then apply risk neutrality, or should we still insist on an increased negative moral weight of evil in the form of simple progressiveness? That is, should the relative progressiveness approach be pure or mixed with some other progressiveness approach?

As mentioned above, what this choice ultimately boils down to is a choice between different degrees of precaution. If we go for pure relative progressiveness, possible harmful upshots are assigned extra negative moral weight only if one of the options would secure a decent mix of risks and chances. This makes for a higher degree of precaution compared to risk neutrality. However, this degree would be even higher should we apply simple progressiveness in our assessment of the moral importance of harmful upshots in the case where no option would secure a decent or acceptable mix of risks and chances.

In addition to this, the mixed progressiveness approach can in turn be constructed in two different ways, one of which makes for a higher degree of precaution than the other. On one interpretation, we distinguish between what we may call an *initial* or *basic* assignation of moral importance to the possible harmful upshots of a decision (according to the simple progressiveness approach independently of what risks and chances would be effected by other options), and an *additional* such assignation (made according to the relative progressiveness approach) in the case where one option would secure an acceptable mix of risks and chances. That is, possible harmful upshots are *always* assigned extra negative moral weights, but when one option secures an acceptable mix of risks and chances, the possible harmful upshots of the other options are assigned *even more* extra negative moral weight. This can be contrasted with the idea of simple progressiveness, rather than governing a basic assignation of moral importance to harmful upshots that will be added to in the presence of an option that secures an acceptable mix of risks and chances, coming into play only as a *complement* in the absence of such an option. That is, the assignations of negative moral weight to harms made on the basis of simple and relative progressiveness respectively are seen as *complementary* rather than *additive*. In the choice between these two versions, it is clear that the additive variant would prescribe a higher degree of precaution compared to the complementary variant. This since, among other things, on the additive variant, the harmful upshots of an option that secures an acceptable mix of risks and chances would be assigned extra negative moral weights that would not be assigned to them by the complementary variant.

On the question of which one of these alternative suggestions – the pure or mixed approach and, in case of the latter, the additive or the complementary variant – can be best supported, I again find it very difficult to come up with a clear answer. On the one hand, the basic moral sentiment that makes me sympathetically inclined towards the general idea of assigning extra negative moral weight to evil also makes me inclined to go for some variant of the mixed approach. On the other hand, however, the reasons that have made me support relative progressiveness rather point me in the direction of applying this approach in its pure form. To this can be added that I am presently completely unable to make a choice between the additive and the complementary variants of the mixed approach, should I consider only the reason I see for accepting a mixed version of relative progressiveness.

Again, this inability to find sufficient reasons to accept one particular version of relative progressiveness, may be seen as a weakness of this whole approach. However, by now, the reader should be familiar with my response to such criticism. I firmly believe that further reasons can be found. The search for these reasons would require a much deeper, detailed and systematic investigation of our moral sentiments in response to risk impositions, however. Presumably, it would require us to consider a vast amount of rather detailed cases where the different versions of relative progressiveness differ in their judgements of the degree of responsibility of the options involved. Again, rather than taking this to be a weakness of the idea of relative progressiveness, my view is to see it as an exciting and promising area for further research. Alas, however, that research would necessitate a wholly different book than the present one.

5.6.5 What Makes for an Acceptable Mix of Risks and Chances?

A key notion in the idea of relative progressiveness is that of an acceptable or decent mix of risks and chances. If, in a situation of choice, one of the options open to the choosing party would secure such a mix, the harmful upshots of the other options are assigned additional negative moral weights. If not, no such additional assignation is warranted (although, if we opt for some variant of mixed relative progressiveness, some additional such weights are assigned anyway). Obviously, therefore, it is of tremendous importance how such an acceptable mix is to be characterised and, even more importantly, on what basis such a characterisation may be undertaken.

In the next chapter, I will illustrate the idea that also the notion of an acceptable or decent mix of risks and chances should be relativised to what is at stake in particular situations of choice, at least to some extent. Compare, for example, the following two cases:

> *Case 1*
> Half of the people living in the world have a life expectancy of 40 years and their lives are burdened by disease, shortage of food and other hardships. However, the other half has been blessed by luck to have much brighter prospects: their life expectancy is about 80 years, they have been able to master most of the hardships suffered by the first half and they are able to enjoy considerable material affluence. Now, this latter group has found a way to extend their life-expectancy even further – to 120 years, say. The resources that would be used to accomplish this change could instead be used for alleviating some of the worst hardships of the destitute group and increase their life expectancy with 20 years.

> *Case 2*
> Half of the people living in the world have a life expectancy of 80 years and their lives are burdened by some hardships. However, the other half has been blessed by luck to have much brighter prospects: their life expectancy is about 120 years, they have been able to master most of the hardships suffered by the first half and they are able to enjoy even more material affluence. Now, this latter group has found a way to extend their life-expectancy even further – to 180 years, say. The resources that would be used to accomplish this change could instead be used for applying the methods mentioned in case 1 in order to alleviate some of the worst hardships of the worse-off group and increase their life expectancy with 40 years.[45]

In case 1, it seems plausible to say that, given the framework of the total amount of resources available, for the affluent group to use the resources at hand for improving their own situation would effect that they enjoy a mix of risks and chances that is far beyond what decency requires. In effect, if the methods that would be used for accomplishing this result would also bring some risks, the moral importance of these would have to be adjusted according to the relative progressiveness idea. However, looking at case 2, it is not equally obvious that effecting a 40 years increase of

[45]It is important that none of the inequalities in the two cases are results of illegitimate transactions having taken place between the groups. Otherwise, considerations of justice requiring one of the groups to compensate the other might be applicable. This, of course, is a difference to the global inequalities of the actual world.

life expectancy for people who already enjoy 80 years of life on average would go beyond a decent mix of risks and chances, at least not to the same extent as in case 1. This since the alternative in case 2 is not to benefit some worse-off group of people, but to benefit people who are even more affluent. In consequence, the case for applying the relative progressiveness idea to the possible risks of the methods that would be used to benefit the first group in case 2 would seem to be considerably weaker that when we consider exactly the same method in the framework of case 1. At the very least, it seems intuitively appealing to suggest that if these risks are to be accorded an extra negative weight, this weight should be smaller in case 2 than in case 1.

At the same time, it should be observed that there is an obvious tension between this way of arguing and other intuitions underlying the relative progressiveness idea. For if we allow the just described idea of relativity of the notion of an acceptable or decent mix of risks and chances to be the only factor determining this mix, every situation of choice will have at least one option securing such a mix. This result, in turn, may seem to contradict the intuitions that, first, in some situations of choice, no option would secure even a decent mix of risks and chances and, second, in other situations of choice, all options would effect outcomes that go beyond what is required to reach such a mix. Consider, for example the case where the life expectancy of all humans has been raised to 300 years and we have the options of retaining this or to raise it to either 400 or 500 years. Here, it may be argued, the mix of risks and chances enjoyed if we do nothing is already beyond the level required for acceptability and decency – and even more so if we look at the improvements that are contemplated. In effect, the relative progressiveness idea should be applied to all risks in such a situation of choice. Similarly, consider the case where all of humanity have a life expectancy of 30 years and the options are to raise this to 33 or 35. It may seem plausible to say that none of these situations would mean that people enjoy a decent or acceptable mix of risks and chances.

For this reason, it may be suggested that, although relativisation to what is at stake in particular situations of choice should have some influence on what is to be counted as an acceptable or decent mix of risks and chances, this should not be the only factor determining this. In addition, we must also employ non-relative ideas. Obviously, this further complicates the question of where to locate the acceptable or decent mix of risks and chances.

On the basis of what has been argued above, my view of this matter should be rather predictable. The ultimate issue regarding any suggestion of what makes for a responsible risk imposition is whether or not this suggestion prescribes a proper degree (and price) of precaution. Therefore, the choice of characterisation of what makes for an acceptable or decent mix of risks and chances should be guided by basic moral convictions regarding this. Again, it seems clear that further and much more detailed normative investigations are needed in order to accomplish this task. However, in the case at hand, the situation seems to be even more complicated than regarding the unclarities mentioned earlier.

If it takes very little for an option to secure an acceptable or decent mix of risks and chances, this will mean two things. First, more situations of choice will be suitable for the assignation of extra negative moral weights according to the relative progressiveness approach. Second, in each such situation, it is more likely that several options may escape such assignation due to the fact that they all secure an acceptable or decent mix of risks and chances. If, contrary to this, the requirements for qualifying as an acceptable mix of risks and chances are very strong, the reverse seems to hold: The assignation of extra negative moral weights according to the relative progressiveness approach will be applicable to much fewer situations of choice while, in those situations where it is so applicable, it is more likely that only one option qualifies as securing an acceptable or decent mix of risks and chances.

On the basis of this elementary observation, we may infer that it is a bit ambiguous how the choice of characterisation of an acceptable or decent mix of risks and chances should be related to the question of what degree of precaution is prescribed. On the one hand, if relative progressiveness is applicable to more situations of choice, this would seem to make for a higher degree of precaution. This since a greater number of decisions will have to involve the assignation of additional negative moral weights to the possible harmful upshots of some of the options actualised in these decisions. On the other hand, however, if more options in each situation of choice to which relative progressiveness is applicable are likely to qualify for effecting an acceptable or decent mix of risks and chances, fewer of these options are likely to qualify for assignation of *additional* negative moral weights to their respective possible harmful upshots.

In this way, then, even if we were to find a criterion of what makes for a proper degree of precaution potent of resolving the other unclarities mentioned above, it remains unclear how the factors of the number of situations to which relative progressiveness is applicable and the number of possible harmful upshots (actualised by the options in each situation) to which additional negative moral weights are assignable should be balanced against each other. In spite of this, however, I remain optimistic to the prospect of further normative inquiry resolving also this peculiar issue.

5.7 Summing Up

I have tried to outline and defend a theory of the moral responsibility of imposing risks built on the basic idea that such a theory provides a unique type of normative reasons that are capable of guiding as well as justifying particular decisions. The most basic feature of this theory is that it is gradual in nature – i.e., it assigns degrees of moral responsibility on the basis of a comparative evaluation of the risks and chances effected by the options open to the choosing party in a situation of choice.

The basis of this comparison is, moreover, an evaluation of those possible harms and benefits produced by the options balanced by their respective likelihoods. This also regards options meaning that some activity is postponed while more information and knowledge is sought in order to improve the quality of the evidence forming the decisional basis. As argued in Chapter 3, it cannot be known how such an option will score in terms of the extent to which it will affect future decision making in a positive or negative way. However, I have argued that the improvement of the quality of available evidence should be viewed as a morally positive factor by itself, a factor that may be balanced against the various costs associated with this type of option.

On the basis of this elementary structure, I have then discussed whether or not risks and chances should be evaluated on the basis of the position of risk neutrality. The alternative is to accept some idea according to which we should be prepared to pay a higher price of precaution (in terms of lost benefits and actual costs) in order to escape certain risks than what would be prescribed by the position of risk neutrality. In the defence of such a suggestion, I have presented a theory of the extra negative moral weight of evil. In the end, I have settled for a version of this idea that I have called relative progressiveness. I have argued that this idea satisfies important basic moral convictions regarding the circumstances in which it may be justifiable to impose risks, while still paying heed to the pragmatic context of the idea of PP.

A by-product of the results of this chapter of some interest to moral philosophers is that the notion of basic moral *rights* does not fit well as a basis for a sound ethical theory about risk impositions.[46] The rejection of the forbidden risks approach wields out all ideas reminding of absolute rights to this effect. This could be compatible with keeping within the structure of a rights-based ethics if, either, the theory limits how a risk imposition may be traded off against chances and other risk impositions so that a collection of smaller chances or risk impositions can never outweigh a single larger risk imposition (a feature that typically sets rights-based ethical theories that allow trade-offs apart from consequentialist ethical approaches), or the theory at least included the element of *waiving* – i.e., the capacity of a recipient of a risk imposition to cancel its moral importance just by approving being exposed to the risk. My rejection of the rigidity of aggregation approach in favour of the relative progressiveness approach rules out the first of these possibilities. The last opening of waiving is similarly ruled out by the fact that the morally relevant factors identified in this chapter may remain so relevant even in the face of a recipient of a risk that approves being exposed to this risk.

To be sure, the idea of relative progressiveness as outlined in this chapter is afflicted by a number of unclarities that may be seen as serious problems. In spite of all these, however, I believe that the framework of relative progressiveness provides a morally sound basis for approaching the issue of what degree of precaution to employ in various circumstances of risky choice. In effect, this idea also helps to specify what price of precaution is the proper one in such circumstances. The various unclarities that remain should therefore be seen as exciting areas for further research in the ethics of risk.

---

[46]Thus, I reject Hansson's suggestion (Hansson 2009) that formulating a theory of a *prima facie* or *defeasible* right against risk impositions is a central task for the ethics of risk.

Given the general approach to the question of the clarification of PP adopted in this book, it follows from what has just been said that we have a specific moral reason to choose according to a version of the requirement of precaution (and, in practice, use a version of PP that implements the prescriptions of this version of the requirement) that prescribes the proper degree and price of precaution according to the relative progressiveness approach. What this may imply in more practical terms will be the subject of the next and final chapter of this book.

References

Allhoff F. "Risk, Precaution, and Emerging Technologies." *Studies in Ethics, Law, and Technology* 3 (2) (2009). doi: 10.2202./1941-6008.1078.

Arrhenius G. and K. Bykvist. *Future Generations and Interpersonal Compensations: Moral Aspects of Energy Use*. Uppsala: Department of Philosophy, Uppsala University, 1995.

Bentham J. *An Introduction to the Principles of Morals and Legislation*. Oxford: Clarendon Press, 1907.

Hansson S.O. "An Agenda for the Ethics of Risk." In *The Ethics of Technological Risk*, edited by L. Asveld and S. Roeser. London & Sterling, VA: Earthscan, 2009.

Hansson S.O. "Adjusting Scientific Practice to the Precautionary Principle." *Human and Ecological Risk Assessment* 5 (5) (1999): 909–21.

Mill J.S. *Utilitarianism*. New York: Bantam Books, 1993.

Munthe C. *Etiska aspekter på jordbruk* (ethical aspects of agriculture). Jönköping: Statens Jordbruksverk, 1997.

Munthe C. *Pure Selection: The Ethics of Preimplantation Genetic Diagnosis and Choosing Children without Abortion*. Göteborg: Acta Universitatis Gothoburgensis, 1999c.

Parfit D. *Reasons and Persons*, 2nd printing. Oxford: Clarendon Press, 1984.

Peterson M. "The Precautionary Principle is Incoherent." *Risk Analysis* 26 (3) (2006): 595–601.

Rawls J. *A Theory of Justice*. Oxford: Oxford University Press, 1971.

Sandin P. *Better Safe than Sorry: Applying Philosophical Methods to the Debate on Risk and the Precautionary Principle*. Stockholm: Royal Institute of Technology, 2004.

Sandin P. "Common-sense Precaution and Varieties of the Precautionary Principle." In *Risk: Philosophical Perspectives*, edited by T. Lewens. London: Routledge, 2007.

Sandin P. "The Precautionary Principle and the Concept of Precaution." *Environmental Values* 13 (4) (2005a): 461–75.

Sandin P. and S.-O. Hansson. "The Default Value Approach to the Precautionary Principle." *Human and Ecological Risk Assessment* 8 (3) (2002): 463–71.

Singer P. ed. *Ethics*. Oxford: Oxford University Press, 1994.

Chapter 6
Practical Applications

We have now reached the point of inquiry where it is time to relate the various theoretical lines of reasoning presented in earlier chapters to actual practices of making risky decisions. In particular, we need to reconsider the issue about what PP should amount to in actual political decision making regarding environmental action and the use and introduction of technology. In doing this, I will proceed in a sort of 'down-top' order, starting with the issue of what the theory developed in the preceding chapter might say about a number of particular activities in isolation from the issue of policy. After this, the discussion will move on to the primary policy level and discuss what my theory implies regarding PP and its implementation within the political and regulative systems of a nation. As a final, I will then conclude by considering a number of implications on a more overarching global and long-term level of policy making.

6.1 General Cases

In this section, the tour of practical application will be started by taking a closer look at our actual consumption and production of goods. It is here that we find the ultimate source of the production of risks and, I will claim, the theory of the morality of precaution developed in the preceding chapter proves to be quite radical in its implications regarding many present practices in this area. Starting out by analysing the general phenomenon of the consumerist society, I will make this point in general terms and, at the same time, underscore the limits set by the theory to this type of civilisation critique. What will unfold is that the theory lends support to rather strong claims regarding the desirability of a substantial global redistribution of wealth. On this basis, I will then proceed to the case of individual consumption and argue that the degree of moral responsibility assignable to the joint production of risks in consumerist societies on the individual level is insignificant for all practical purposes. This takes us to the larger scales of mass production, where it is concluded that while it seems to be much easier to assign a significant level of responsibility for our unjustifiable production of risks at this level, it is not the case that all of the blame can be shifted in that direction. Together, these different conclusions jointly

C. Munthe, *The Price of Precaution and the Ethics of Risk*, The International Library of Ethics, Law and Technology 6, DOI 10.1007/978-94-007-1330-7_6,

form a strong basis for arguing that regulative policy rather than moral blame, self-discipline or education is the important outcome of the theory of the morality of precaution. Before turning to the questions regarding the design of such policies, however, I will consider a number of hard cases where it is less clear what the theory of the morality of precaution has to say and point to the challenges created by this practical indeterminacy on the level of policy making.

6.1.1 Consumerism

Myself and, I presume, the vast majority of my readers live in a society where we are being constantly bombarded by the introductions of new brands, gadgets and products. For most of these it holds that the wants or needs that they are uniquely made to satisfy are very few and weak, if at all existent. Brand *X* or *Y* or *Z* will, for the overwhelming majority of cases, do just as well as any other version of a particular product. Last year's model is, in all aspects that matter, just as good as this year's upgrade or new sensation. In many cases, whole product-families seem to be equally superfluous from the perspective of how important they are for satisfying significant wants and needs. This is the world of *consumerism* – an economic system built on the continuous consumption of new products, thereby promising to secure (directly or indirectly) a steadily increasing quality of life for those being a part of this system.[1] This is a system of which I myself am a part, and, I must confess, which I personally enjoy very much!

Nevertheless, from the point of view of precaution, consumerism seems to be open to criticism. This critique builds on four observations: First, the promise of a steadily increasing quality of life presupposes that consumerism does not have side-effects that will undercut this process. Second, one type of effects in particular that would effect such an undercutting is the destruction of environmental prerequisites for the life and well-being of humans. Third, many serious environmental threats seem to be the effects of consumerism. Fourth, and perhaps most important, the fuel that keeps the engine of consumerism running is the steady introduction of new types merchandise and services, the marginal value of which is extremely small.

Together, these four aspects of consumerism strongly suggest that the basic mechanism of this socio-economic culture – the continuous introduction and consumption of new goods – can be criticised on the basis that each step in this process necessitates the imposition of unnecessary risks. Each (or nearly each) new product introduced on the market will produce extra risks, mostly for the environment but

[1] In Western consumerist societies of today, I would suggest, this growth is mainly indirectly effected by actual consumption (through the use of resources created by the consumerist system for other purposes, such as health care and public health, education, social security, public safety, et cetera). This since people in these societies as a rule live on such a high level of material wealth that the marginal value of each new brand or product may be assumed to be microscopically small. However, in other consumerist societies, such as developing countries, the continuous consumption of new goods may still by itself be an important contributor to many people's quality of life.

also for human life and well-being in its own right. And even if these extra risks are not so great, the needs that the new products may satisfy are almost never even close to being of the magnitude or importance required for balancing the extra risks. In other words, we could abstain from each new product and thereby avoid the extra risks without having to sacrifice anything of significant importance.

True, in some isolated (and relatively unusual) instances, the new product will serve to alleviate or reduce risks imposed by other products. Consider, for example, a new formula for paint used by professional painters that contains a solution agent that is significantly less toxic than that of the old formula. Or take the case of new products the production of which requires the use of less energy than preceding products of the same type (thereby economically benefitting the producers, but also reducing environmental risks due to energy production). In these cases, it may seem less unlikely that the introduction of the new product may be justifiable from the point of view of precaution. However, whether or not such a justification would in fact be forthcoming also depends on further factors, such as whether or not the new product, besides reducing old risks, also introduces new ones,[2] what body of evidence there exist in support of the assessment of risks and benefits and what would be the cost in terms of actual harm, prevailing risks and lost benefits of either improving this body of evidence or abstain from using this particular type of product at all. For many types of products it seems to hold that we could actually manage very well without them altogether, or at least for a very long time. Consequently, even if a particular product would reduce the risks of other products and on this basis be defendable from the point of view of precaution, it does not follow that the use and production of these *other* products may be so defended in the first place.

This is not the end of the story regarding consumerism and precaution, however. At the same time as the just outlined reason for criticising consumerism on the basis of precaution is indeed valid, it has the weakness of resting entirely on the observation of the very small marginal value of each product or type of product taken in isolation. However, this way of analysing the situation misses out on what seems to be the main benefit of consumerism – namely, the aggregated effect of the *totality* of this system on our level of well-being. This effect – what economists refer to as *growth* – is what enables all of us to be in the position that we may remain on a decent level of well-being even if we were to abstain from the latest model of this or that product or this or that new type of gadget. This since growth is what pays for our consumption of more vital kinds of goods, such as nutritious food, decent housing and clothing, health care services, et cetera. In effect, while particular aspects of consumerism taken in isolation may easily be criticised for introducing unnecessary risks, or risks which we could easily abstain from creating without loosing out on a decent level of well-being, the same cannot as easily be said about consumerism as a whole.

[2]For example, in the case of less energy consuming products, the end result may be even larger energy consumption than before due to an increased market for the product created by lower prices made possible by the energy saving production process.

Seen from this perspective, the idea of the modern Western-style society as a 'risk society' is a misnomer.[3] For it seems to be an undeniable fact that the members of developed consumerist societies suffer a much less severe mass of risks to life, limb and quality of life than do people in less developed and less consumerist societies. True, it also seems that the members of consumerist societies are more prone to note and worry about the risks they are indeed exposed to. However, that has less to do with the risks actually present than with quite understandable human psychological mechanisms. Perhaps it also, to some extent, reflects the type of moral judgements employed in the former chapter in support of my basic theory of the morality of precaution. In any case, a notion of the 'risk *sensitive* society' would be more apt to capture such aspects.

While this may be so, it is still an open question to what extent the *actual* content of consumerism as it appears in developed countries can be defended from the point of view of precaution. For while it may be true that consumerism as a whole is a prerequisite for securing and upholding a decent level of well-being for great masses of people, it is still an open question whether or not the joint risks produced by this practice are too serious for the aggregated mix of risks and chances of consumerism to reside on an acceptable level. This becomes especially clear if we consider the possibility of securing a decent level of well-being through consumerism (albeit, possibly, a lower level than what is presently enjoyed in consumerist societies) without producing as serious risks as what is presently the case.

This possibility reveals itself through the observation made above that the problem of consumerism from the point of view of precaution seems to be the very slight marginal value of each step in the consumerist process of the continuous introduction of new products. However, if we compare each such step not with the immediately preceding situation, but with a much earlier situation, this value may appear much more significant. Take, for instance, the case of the latest model of mobile phones. Compared to the preceding model, this step may seem to be of almost no marginal value – we would all have managed equally well if the new model had never appeared on the market (maybe even better, if we find the finesses of the new model mainly to create unnecessary complications in our lives and buy it mainly as a result of social pressure). However, compared to the much earlier situation when we did not even have access to traditional telephones, the latest model of mobile phones may seem to constitute a significant improvement of the human condition. In addition, the extra risks created by the presence of this particular product in the market compared to the situation when no phones at all existed may seem to be clearly balanced by the benefits of having access to the latest model of mobile phones rather than no phones at all. Especially so, since the absence of phones altogether gave rise to quite significant risks, for example, to health due to the problems of getting in touch with a physician in an acute case of disease.

[3] This notion was introduced by German sociologist Ulrich Beck (1992).

This complexity of the situation suggests that the problem of consumerism from the point of view of precaution mainly has to do with its *pace* – the lack of temporal space between the different steps in the consumerist process of the continuous introduction of new products. For if, instead of immediately introducing a particular product, we were to wait and keep investigating to what extent the new product really constitutes any significant improvement and what possible risks it may create, we have the opportunity of using these investigations as a basis for further developments of the product in question, designed to increase its marginal value and to reduce its possible risks.

Obviously, there is a limit to how much of 'space' between the steps in the consumerist process that may be justified in this way on the basis of the requirement of precaution. Holding on to the example of telephones, we may find that the step from no phones at all to the first working telephone networks could indeed be justified in spite of the case that the space between these two steps is rather tight – this due to the obvious benefits and risk-reductions brought about by the access to phones among many people. Due to this, the cost of postponing the introduction of phones pending further investigations of their risks and benefits would, it may be argued, become intolerably high. That is, such a postponing would be the expression of an unacceptably high price of precaution. However, it seems much more difficult to justify several of the succeeding steps in the development of telephones and telephone networks compared to the step when telephones were introduced for the first time. This due to the simple fact that the real substantial benefit of phones (the possibility of instant and interactive communication at a distance) has already been secured, at the same time as each new developmental step beyond that creates additional risks (at least if we assume these new products to be mass-marketed). In effect, the cost of postponing the introduction of new phones seems easier to justify, since we may argue that it is not worth creating extra unclear risks when we can manage almost as well without creating them.

To this it may be retorted that applying this type of reasoning to consumerism as a whole would lead to a situation where society would not be able to secure an acceptable level of well-being. This since not only the width (in the sense of the number of areas of human life for which there exist products on the market), but also the pace of consumerism is a critical factor in its ability of creating a minimally decent basis of wealth for everyone. I agree to the first part of this suggestion – the pace of consumerism indeed is important for the production of wealth. However, this does not imply, that *any* increase of the pace is able to create enough of additional wealth in order outweigh the reasons for slowing down the pace in order to reduce risks.

In conclusion, the theory of morally responsible risk imposition developed in this book and the interpretation of the requirement of precaution that can be made on the basis of this theory strongly suggest that the consumerism actually at work in developed countries indeed involves quite a lot of morally irresponsible risk impositions. Thus, on the whole, it would be desirable if a higher degree of precaution was exercised – something that would seem to imply a demand for a slower pace of consumerism. In particular, it seems to follow, first, that in the choice between different products (or that of having a particular type of product or not) where the difference in positive value (i.e., what benefits they may bring) is negligible but where there is reason to believe that one of the choices could produce additional risks, we have reason to postpone the possible introduction of this product in order to improve our evidence regarding the risk-profiles of our options and to further develop the product in question (unless, of course, this would bring unacceptable costs in other respects). However, also in cases where the product in question would in fact bring some benefits, we have to accept that these may not be substantial enough to balance the extra risks created by having this product on the market.

Now, all of this presupposes a context of developed nations where it can be reasonably claimed that people, in most cases, can do without the new product without having to sacrifice anything of substantial importance. However, as we all know, things are different elsewhere on our planet. In some parts of the world there may be sufficient wealth available for securing an acceptable level of well-being for everyone without the creation of extra risks, but the actual patterns of distribution impede such a result. In these cases, it seems fairly obvious that accelerating the growth by introducing a lot of cheap products that may benefit those that are worse off to the price of creating a lot of extra risks would not be a defensible policy.[4] This due to the simple fact that benefiting those that are worse off may as well be secured by changing the distributive policies. To be sure, this improvement for those worse off will then be achieved at the expense of those better off. However, since the latter will remain safely above the threshold for an acceptable level of well-being and since the creation of additional risks will be avoided, this type of measure would still seem to be clearly preferable from the point of view of precaution.[5]

In many cases, however, the amount of wealth actually available in a country or region would not suffice to secure an acceptable level of well-being for everyone even if it was to be distributed more equal. In these cases, the theory of morally responsible risk impositions developed in the preceding chapter, clearly allows for more of extra risks being created in order to achieve a level of growth that may lead to a situation where everyone may be guaranteed an acceptable level of well-being. First, since people enjoy a lesser quality of life, chances are much better that any improvement of their well-being through the consumption of material goods will also be a significant improvement – i.e., we may assume additional products introduced on the market to be much more likely to have a substantial positive marginal value.[6] Second, since people have not yet secured an acceptable or decent level of

---

[4]Unless, of course, there are reasons for believing that the growth itself would change the political situation into one where more people are allowed to enjoy more well-being up to a point that may justify the extra risks.

[5]This presupposes the absence of dynamic effects in the form of negative growth, for example, due to those better off in such a society becoming motivated to emigrate (personally and/or financially). Below, such prospects will be noted as one of many factors supporting the idea that precautionary policies should ideally be at work on a global rather than merely multi-lateral, national or regional level.

[6]Assuming, of course, that they are to any benefit at all.

risks and chances, the price and degree of precaution that can be justified is much less than if the opposite had been the case. In effect, from a moral point of view, it takes less of extra chances of benefits in order to balance the creation of some particular risk.

In Chapter 4, I mentioned that one aspect to receive some attention in the development of an ethics of risk is considerations of justice. In that context, I expressed some cautious doubt as to whether or not such considerations should really be seen as a basis for or parts of the morality of *precaution*, but suggested that such a morality may nevertheless support ideas in line with a theory of justice. What we now see, is that the requirement of precaution, at the very least, seems to support claims to a more equal global distribution of beneficial risk production. Somewhat simplified, what has just been demonstrated is that while developed countries are overproducing risks, less developed countries are in fact *under*producing risks. In effect, a global redistribution of risk production from more developed to less developed countries seems called for from the point of view of precaution.[7] This, of course, assuming that we are talking about risks that are not unnecessary, but would actually be accompanied by sufficiently beneficial effects.

Later on in this chapter, I will return to this implication and discuss its long-term consequences for global environmental policy areas, such as climate change. However, for such a discussion to be possible, we must first consider what should be the more precise target of precautionary policies, what function they should be designed to have, on what societal level such policies should operate and, finally, what should be their actual content.

6.1.2 Why Individual Motivation Should Not Be the Target

One part of the mechanism that keeps the consumerist clock ticking is, of course, the actual consumer behaviour of individual people. Moreover, this particular part seems intimately connected to the production of risks within consumerist societies. The overproduction of risks in developed countries pointed out above had not been possible without certain patterns of consumer behaviour among people and the same holds regarding the underproduction of risks in less economically and technologically developed regions of the world. Had the poor people of the world consumed according to a pattern like those in developed countries, these people would also have produced more of technological and environmental risks (as well as well-being to be enjoyed by themselves, of course).

This may be seen as a reason for trying to extract the important implications of a theory of the morality of precaution at the individual motivational level: The analysis of practical application should focus on the moral motivation of individual people – those participating in the overproduction of risks and those capable

[7] Of course, the very distribution may bring risks of its own, which then will have to be taken into account in decisions on its size and shape.

of making it possible for individuals in settings where risks are underproduced to produce more risks than they are currently capable of (often, these people are one and the same). That is, moral education (to make the overproducers embrace the moral reasons for limiting their consumption in order to secure a morally responsible level of risk imposition) and charity (flowing out of the moral motivation of the adequately educated overproducers to help the underproducers to accumulate wealth that will in turn produce more risks) are the chief practical messages of our theory of the morality of precaution. Or, at least, so it may seem.

What this means in ethical terms is that the idea of a virtue of precaution seems to get a second chance. Even if, as I argued in Chapter 4, we have good reason to resist an analysis of the basic morality of precaution along virtue ethical lines, the fostering of precautionary virtues may after all have an important role to play in the practical application of the basic theory.

However, this way of perceiving the situation misses out on an essential part of much of the risk production in consumerist societies briefly touched upon above. For if we take phenomena such as the total mass of risks to the environment produced in our own society, or the total risk level effected by our collective use of some type of technology, these cannot as a rule be traced back to the behaviour of the individual consumer by fractioning down step by step in a neatly linear way the risk-benefit ratio found on the higher level. This due to the well-known phenomenon of synergies in the addition of very many contributions that jointly form a whole larger than the simple arithmetic sum of its parts. In effect, even if we take a phenomenon such as the quite significant environmental risks produced by our joint use of chlorine for washing and house cleaning purposes, we are unable to say about any single individual behaviour in this area that *it* gives rise to any environmental risks *at all*. For in each such case it seems to hold both that had this behaviour existed in isolation, nature would have had no problem dealing with the resulting small amounts of chlorine being emitted, and that if such an isolated individual contribution was to be subtracted from the actual chlorine pollution, this would make no difference to the environmental risks produced by this pollution. In effect, many environmental problems result from individual *non-polluting* behaviours that join up to form a mass of behaviours capable of passing the threshold where pollution results.

This well-known nature of many environmental problems makes it problematic to apply our theory of the morality of precaution at the level of individual consumption in three partly interconnected ways. First, we are deprived of any uncomplicated reasons to assign blame to individual consumers. Each of them may be justified in claiming that the unacceptable risk production of consumerist societies is no fault of theirs. To be true, some writers have suggested that in situations such as these, it may be claimed that each party is responsible for a proportional fraction of the totality of the bad outcome.[8] However, and this is my second point, even if this was to be granted, measures designed to influence the moral motivation of a single individual will still be hard to defend to the extent that they also effect some bad effects for

[8] See, for example, Glover (1986) and Parfit (1984, Chapter 3).

these people (such as having a slightly more complicated life, for example, when cleaning the house). Third, partly as a result of this, it would be very difficult to make anyone see any point in such measures and, in effect, to gain political support for policies based on the idea of blaming the individual consumer for the unacceptable risks produced by consumerist societies.

6.1.3 Precaution as a Collective Good and the Need for a Politics of Power

From this it may be tempting to infer that defensible precautionary policies should concentrate only on the motives of those single individuals in consumerist societies that actually do produce significant risks. For besides individual consumers, consumerist societies also keep ticking thanks to the larger apparatus of production and distribution and on this level we may often be able to trace significant risks directly to individual contributions. The motives and resulting behaviour of any of the large petro-chemical corporations (to take just one example) does produce quite significant risks even if we consider it in isolation from other producers. In effect, the argument goes, it is towards such acting parties that the force of virtue based precautionary measures should be directed.

I concede the point that it may seem rather more easy to justify precautionary measures along the virtue ethical lines presently under consideration in the just outlined way. However, it is a mistake to believe that the levels of, on the one hand, consumption and, on the other, production and distribution can be kept separate to the extent required by this line of argument.

First, obviously, any measure succeeding in influencing the motivation and behaviour of a producer of some type of merchandise in a direction desirable from the point of view of precaution will necessarily affect consumers (for example, through higher prices, reduced quality and a slower flow of new things to deliver that charm novelty we apparently are so easily seduced by). This effect would be blocked only if the gap in the market created by the change of behaviour of this producer would be (almost) immediately filled by another producer (who thereby would gain market shares). Moreover, regardless of how effective a system for education and/or indoctrination was to be set up in order to foster the appropriate precautionary motives, we may rest assured that this is exactly what would happen.[9] Thus, precautionary measures working at the level of individual moral education would be pointless from a practical point of view.

Second, I am prepared to claim that it would be rather disturbing from a moral point of view to choose the target and function of precautionary measures only according to a principle of individual blame. As was underlined above, the most

[9] This claim does not assume that any *conceivable* market gap will be automatically filled, only that gaps resulting from measures that impede a formerly active part of the market are very likely to be filled.

troubling aspects of our production of risks are not the results of individual wickedness or neglect – be it on the motivational or purely behavioural level. Rather, it is the faulty pattern of coordination of the totality of individual behaviours that should be blamed. In that light, the idea of directing *all* force of precautionary measures at only those individual members of the collective where a pattern of individual blame can also be detected seems to me to be both unfair and ineffective. Especially so since this strategy is motivated only by the idea of casting all precautionary measures in a virtue ethical mould.

Two conclusions seem to me to flow from this discussion. First, appropriate precautionary measures need to target not only, or even primarily, individual acting parties, but rather collective behavioural patterns that coordinate individual behaviours into those aggregated collective behaviours responsible for the overproduction of risks. Precaution, to use an economist term, is mainly a *collective* (or public) rather than an individual good. Second, these patterns cannot feasibly be attacked primarily with ideological or educational means, but need a politics of power to produce the desirable result. Several hundred years ago, British philosopher Thomas Hobbes realised, when pondering what would happen in his imagined state of nature of "everyone's war against everyone" if some people where to lay down their swords out of the conviction that peace would be a more desirable state of the world,[10] that any qualitative change out of the situation would need the force of raw power: a 'Leviathan' (or, as we call it today, the state or government) that by force and threat is able to change the prerequisites of society so as to make violent action unattractive for individual acting parties.[11] In a similar manner, if the ideal suggested by the theory of the morality of precaution set forth in this book is to be actualised (or even approximated), we need political measures that are capable of making whole masses of individual acting parties to change their ways. That is, we need policies that go way beyond influencing the moral motivation of people and works directly at the behavioural level by making adherence to such behavioural patterns that would coordinate individual behaviours into collective patterns desirable from the point of view of precaution attractive also for those who are not moved to a sufficient degree by the moral reasons set forth above.

True, if such precautionary policies were to be implemented, this might eventually influence the moral motivation of people so that they come to embrace the idea of doing their part in this implementation to a greater extent. If so, this would make the enforcement of such policies more cheap, easier and less in need of elements of coercive pressure and would, for this reason, be a much welcome side-effect. Indeed, this may be seen as a strong reason for combining the more outright forceful measures of a politics of power with strong elements of education, indoctrination and propaganda in order to boost such a development. Moreover, it may be a reason for some moderation in the use of force in order to make people adhere to the precautionary ideal, since the coercive pressures may otherwise alienate people

---

[10]They would simply be slaughtered, held in slavery or at least robbed of all their possessions.

[11]Hobbes (1651).

from this ideal to an extent that would impede its implementation in the long run. However, none of this contradicts the main point that precaution is a collective good and that, therefore, precautionary policies cannot be limited to measures designed to stimulate the creation and cultivation of precautionary virtues.

6.2 Hard Cases

Much as one may concede the claims about the problems of consumerism from the point of view of precaution, the implications of these regarding issues of justice and the need for a politics of power to deal with these issues, there remains the question of what actual content such policies should have. Above, I have argued that many activities and practices in developed societies are part of a significant overproduction of risks undertaken at the expense of people in other parts of the world. From the point of view of the morality of precaution, this constitutes a powerful reason for implementing policies capable of modifying cultures, economies and life-styles of developed societies in a way that would facilitate a transfer of beneficial risk-production to less developed parts of the world. However, this general claim leaves open exactly what parts of consumerist societies should be so modified and to what extent this should be done. In the next two main sections of this chapter, I will discuss the formal aspects of such policies. However, before that, I will discuss three major areas of consumerist societies, where it seems especially difficult to assess the exact implications of the theory of the morality of precaution set forth in this book: pollution such as the sort effecting climate change, nuclear power and biotechnology.

6.2.1 Climate Change and Pollution

Industrial facilities for production, especially of energy, and motorised vehicles for transportation emission vast amounts of environmentally damaging substances into the atmosphere, oceans and the earth. These being interconnected natural systems, it is nowadays generally agreed that this practice constitutes *the* major environmental threat on a global scale, and also includes patterns of food production (methane emissions due to meat farming). A prime example has become the emission into the atmosphere of so-called green house gases and the nowadays generally acknowledged effect of this for the climate through the phenomenon of so-called global warming. This phenomenon brings terrible risks for human life and well-being even under the most cautious scenarios (especially so if socio-economic side-effects are weighted into the balance).[12] In the following, green house gas emissions and

[12]For facts on this issue, see IPCC (2007). The side-effects include, among other things, massive migration from large areas that become uninhabitable due to climate change, severe strain on traditional industries, and resulting risks of grave social unrest and war. It should be observed that it is highly probable that these effects will saliently affect also developed nations.

climate change will be a model template for discussing the application of my idea of the morality of imposing risks and its implied ideal of precaution to pollution in general.

While the problematic aspects of pollution just outlined is generally conceded to,[13] at the same time, it is also recognised that climate change (and most other pollution) is a side-effect of practices that in many ways bring substantial benefits to people: Heating in cold climates, appropriate housing, hygienic infrastructures for sewerage and water supply, efficient production of vital goods such as food and clothing, the ability of swift transportation over large distances, et cetera. This has given rise to the idea of a global upper limit for pollution and the ability of nations to trade their respective shares of 'pollution rights'. Regarding the emission of green house gases, such an idea was formalised in the so-called Kyoto agreement of 1997.[14] Since then, the attempt to create an agreement along similar lines but involving all nations and thus being capable of reducing significantly the risks of climate change has been persistent, albeit remarkably slow and quite unsuccessful.

The core of the Kyoto solution may seem congenial to the ideas set out above, since it facilitates the redistribution of risk production in this area from one nation to another with the help of ordinary market mechanisms. Nations that so prefer can sell their pollution rights for other goods or buy additional pollution rights for such goods from other nations, while the total emission of environmentally damaging substances is kept within acceptable limits. Moreover, since the quality of the basis of knowledge used for predicting the consequences of various emission levels becomes exponentially worse the more emission levels are allowed to rise above the previously known,[15] in this case, also the reason to improve the evidential basis of risky decisions speaks directly in favour of imposing a rigid upper limit on global emissions in order to reduce the uncertainty of the risk scenarios underlying further decision making.

However, in order for this strategy to actually promote precautionary ideals, the system has to be set up in a way that makes the market mechanisms actually work in their favour. In order to achieve that, at least two requirements need to be met. First, the global upper limit must be neither too high, nor too low. Second, within this limit, the system has to work from an initial endowment of pollution rights that is appropriate according to the theory of the morality of imposing risks set out above. Both these requirements give rise to peculiar problems.

[13]This claim is consistent with the existence of the so-called climate scepticism, expressed by a minor group of researchers (mostly not in the relevant scientific fields), who doubt the conclusions drawn, e.g. by IPCC. However, none of those doubts disprove the *risks* created by green-house gas emissions, global warming and climate change and – for the most part – concern minor details in the scientific discourse. Some climate sceptics also criticise the *political initiatives* taken to handle these risks, and thus seem to assume some (undeclared) theory on the ethics of risk and precautionary policy. My discussion in this subsection may be seen as a response to such debaters, with the difference that I present and defend a transparent and worked out theory on the ethics of risk and precaution on which the analysis of climate change policy is based.

[14]United Nation's Framework Convention on Climate Change (1997).

[15]Rummukainen (2005) and IPCC (2007).

To start at the end, one of the aspects of the Kyoto agreement that has been the target of much criticism is how it endows the initial pollution rights – that is, 'shares' of the total accepted pollution level. This issue has continued to haunt global negotiations as developed countries have demanded that also developing and outright underdeveloped nations have to pay a part of the price of precaution in the face of climate change by limiting their (future) emissions. These latter countries, in contrast, have argued that their part of the price of 'climate precaution' has to be small enough not to block their prospects of growth and development of the sort already achieved in developed countries. In the reality of negotiations, this basic conflict attains some complexity through various proposed schemes of compensation or aid aimed at influencing the price of precaution actually paid by a nation. In essence, though, the issue is about distributing the burdens of limiting green house gas emission, and may as such be addressed in terms of the initial endowment of pollution rights.

According to my theory of the morality of precaution, this endowment should be made not, as is actually the case to a large extent within the Kyoto protocol, according to the principle 'he who has should receive'. That is, the point of departure should not be a nation's actual production of pollution (or risks) alone, but this production set in relation to what level of wealth it secures, how this level would be affected by different decreases in the production of pollution, how other nations compare in these respects and how all of this relates to the idea of an acceptable mix of risks and chances. It does not seem a very bold conjecture to suggest that, should such an analysis be undertaken, the outcome would be in line with what has been said above about consumerism in general. Hence, set in relation to the pollution rights that *should* be allotted with regard to climate change, developed countries are monstrous overproducers of pollution, while less developed countries are tragically underproducing risks in this area.[16]

Not very surprisingly, an initial endowment based on such an analysis alone has not been accepted by developed countries, especially not the most rich and powerful. Crudely put, the official reason for this stance is the claim that they have a right to maintain a level of wealth that approximates the level which they have managed to secure for themselves – at least as long as the possible reductions of this level is not proportionally reciprocated by less developed nations. This argument obviously ignores the fact that the level of wealth of rich countries is being paid for with a vast overproduction of pollution risks.

[16]This argument must be kept separate from an argument based on desert considerations often wielded by representatives of developing countries. That argument claims that rich countries are overproducers of pollution, while developing countries are not, since it is mainly the past pollution of developed nations that has given rise to the climate change problem now being faced by all nations – simply put: 'it is their fault not ours!'. This claim assumes desert to be a valid normative consideration, but such an assumption is not a part of my theory. Since desert is a classic suggestion in debates about justice, however, this is another instance of how my theory is able to produce results of relevance from the point of view of justice without making any particular assumption about what is a just distribution of risks.

At the same time, from a political standpoint, it seems reasonable to guess that part of the cause of this position is that it would presumably be difficult for any single government of such a country to gain long-term electorate support for a policy that would drastically decrease the level of material wealth enjoyed by its citizens – at least if a reciprocating scheme of the sort just mentioned is not part of the deal. This, in turn, makes it rather likely that, should developed countries after all accept a change in the ratios between different nations as regards the initial endowment of pollution rights more in line with my theory, this would in fact be bought at the expense of an increased global upper limit of pollution that would be clearly too high from the point of view of precaution. Alternatively, the pace of the adjustment to lower global emission levels may be slowed down, so that the risks are allowed to become even more serious, necessitating even more drastic adjustments in the long run. This situation is problematic in several respects.

First, it highlights the problem of how to organise the determination of the global upper limit of allowed risk production and the initial endowment of risk production rights in any system of policy aimed at implementing the ideal of precaution worked out in this book. The example of climate change and pollution trade illustrates this on a global scale. However, in principle, the same problem is apparent also within nations, where powerful acting parties have the ability to direct the political agenda and the outcome of political decisions. Secondly, a tension is made salient between, on the one hand, the aims set forth by the ideal of precaution and, on the other, the liberty of individual acting parties (people, companies, nations etc.) as well as the democratic processes of decision making of individual nations. Third, we see an apparent conflict between short-term pragmatic concerns and more long-term goals of a more idealised nature. The first two of these problems will be returned to later on in this chapter. I will, however, close this sub-section on climate change and pollution trade by commenting on the third problem, which seems to me to be at the heart of this particular controversy.

On the basis of the observations made above, two very different conclusions could be put forward. First, we could adopt a pragmatic stance and claim that we (regretfully) have to learn to live with the inevitabilities of political reality. That is, as things stand, it is acceptable (although not ideal, of course!) from the point of view of precaution to institute a system of pollution trade that is closer to the type that seems to flow out of the Kyoto agreement. Alternatively, a deal more close to the one indicated by my theory of the morality of precaution could be struck between willing parties, at the expense of excluding some of the most important contributors to climate change, such as the USA or China. While far from ideal from a precautionary point of view, these outcomes are still *something* and such a something is, after all, better than nothing. On a more hopeful note, with some luck and hard political work, such flawed systems may with time be made to develop into a system more in tune with the ideal of precaution.[17]

[17] In such a process, it may be claimed that the idea of directing policies towards the fostering of precautionary virtues among the electorate of the nations involved has an important role to play.

Secondly, we may instead adopt a more principled stance and claim that the 'learn to live with'-strategy amounts to surrendering and giving up the precautionary ideal. By doing this, we merely serve to uphold (or even strengthen) the established pattern of over- and underproduction of risks in this area. What is worse, economic forces are likely to lead to a result where the most wealthy nations either will assume the role of free-riders that play the rest for suckers, or buy themselves rights to even more risk production for a rather cheap price (since they are in such a strong position of bargain in relation to less developed countries), which in turn increases the underproduction of risks in the selling countries and, thereby, serves to uphold and strengthen the vastly unfair global inequalities in wealth between nations.[18]

In my view, both of these stances have something to them. Pragmatically motivated compromises are indeed an important political instrument, in particular regarding the pursuit of long term goals far removed from the realities of the present situation. However, at the same time, if compromises are pushed too far, they become counterproductive from the point of view of the long-term goals pursued and then the appropriate response may indeed be refusal to make an agreement. Similarly, when different nations try to push their own agenda in order to have a compromise that is as close to what is in their prudential interest as possible, the risk is obvious that the accomplishment of an actual agreement is delayed so that the problem that the agreement is aimed at handling becomes even more serious.

Regarding the particular case of climate change and pollution trade, the actual situation at the time when this book is written[19] unfortunately seems to reside quite far from an ideal mix of the two strategies of pragmatic adaption and principled idealisation. Such an ideal mix should be shaped in such a way that if all parties were to apply it, a satisfactory agreement would be in reach. In game theoretical terms, at the moment the parties seem to be dug down in the strategy of playing an advanced multi-player version of the *Chicken* game,[20] i.e. prioritising the pursuit of their own relative self-interests, hoping to push the other parties to yield into paying a larger part of that price of precaution that has to be paid by someone if the grim scenarios of climate change are to be avoided or at least ameliorated. As mentioned, in the meantime, these scenarios grow grimmer and our ability to predict the outcome (thus being able to prepare for it) becomes weaker and the joint application of this stance therefore threatens to lock the parties into something close to a genuine multi-player *Prisoners' Dilemma*.[21] In contrast, what would seemingly be needed

[18]This last effect might be thought to be countered by a system for price control working in favour of less affluent countries. However, it seems unlikely that this would be accepted by developed countries, since they could then just as well have accepted to be endowed with less pollution rights from the very beginning, and as a bonus escaped the transition costs of upholding the price control system. Alternatively, these same nations may accept such a system but engineer it to their own benefit, much as has been done in other areas of global trade (cf. Pogge 2008).

[19]About a month after the failure of the COP-16 meeting in Cancún, Mexico, in 2010.

[20]For a basic outline of the game theoretical features of Chicken, see Resnik (1987).

[21]For a basic outline of the game theoretical features of The Prisoners' Dilemma, see Resnik (1987).

is the joint application of a *Tit-for-tat* type strategy,[22] with all parties playing an at least somewhat generous stance as the opening card. I conjecture that an analysis according to my theory of the morality of imposing risks and the resulting version of the ideal of precaution defended would reasonably approximate what might be expected as the eventual outcome of such a more constructive take on the negotiations. Unfortunately, this directions does not seem to be where the political process is heading at the present time.

As a result, a more general point can be made on the basis of the apparent soundness of both the pragmatic and the principled stance, and the apparent present unwillingness of parties to at least apply a tit-for-tat strategy. Regarding issues where much is at stake, such as climate change, this situation may be seen as an argument against the type of bargain and agreement model of policy making that is currently the dominant practice in international politics, and in favour of political power mechanisms, such as majority decisions on laws that are then effectively enforced on individual parties. I will return to this point in the final section of this chapter.

6.2.2 Nuclear Power and Energy Production

A classic example of a risky technology is the production of energy through the process of nuclear fission. Originally celebrated during the 1950s and 1960s as the peaceful and beneficial side of the branch of science that gave us nuclear weapons, nuclear power became increasingly controversial during the 1970s and severely questioned after the so-called Three Mile Island incident. In that case, a nuclear power plant close to the city of Harrisburg on the U.S. east coast malfunctioned in a serious way, making people all over the world conscious of the threat of a nuclear 'melt down'. That is, if the technological means of controlling the nuclear fission process where to fail, the effect may be a massive emission of severely radiating and toxic substances (such as plutonium) in the air, soil, drinking water, et cetera. Luckily, the Three Mile Island hazard never went all the way to such a disastrous conclusion (although significant amounts of radiating substances were emitted into the atmosphere). However, in the 1980s the total collapse of the nuclear power plant of Chernobyl in the then Soviet Union provided a grim illustration of the consequences of a nuclear power hazard brought to its final conclusion. There is no need to go into detail about the monumental suffering brought on people in a large area around Chernobyl or the sustained long distance effects on the environment as far away as Scandinavia. It suffices to note that, when we talk about technological disasters, a nuclear power hazard has become the paradigm of the worst we can imagine.[23]

[22]The importance of this strategy for establishing mutually benefiting cooperative schemes in situations with radically opposed interests has been brilliantly described and analysed by Robert Axelrod (1984).

[23]While making the last corrections to this book, the hazard at the Fukushima nuclear power plants, classified as serious as Chernobyl, was still under way.

Partly as an effect of the growing criticism of nuclear power in the 1970s, attention was also brought to its peculiar *waste* problem. While not emitting any damaging substances directly, if functioning appropriately, nuclear power plants necessarily leave us with waste products that remain extremely dangerous for *very* long time-spans (approximately 100,000 years). There are methods to shorten this time, but such processes at the same time tend to make the waste even more lethal and the time during which it needs to be tightly contained and controlled will still present a major management problem for any society.[24]

At the same time, nuclear power is in some other respects a very attractive source of energy. It facilitates comparably cheap and efficient energy production on a scale several times larger than any alternative technology.[25] Moreover, if a nuclear power plant is appropriately constructed and managed (so as to avoid serious accidents) it does not pose any immediate danger to the environment or health. On the contrary, under such conditions, hazards are very unlikely to occur while, at the same time, a nuclear power plant does not produce any significant amounts of pollution. On a long-term scale there is, of course, the waste problem mentioned above. However, rather than inevitable large-scale pollution of the present environment, this aspect is about possible local emissions sometime far off into a highly uncertain future, the likelihood of which may be greatly decreased by appropriately designed storage methods. This, at least, is what would be claimed by the nuclear power lobby.

Now, from the point of view of precaution, it may appear that even if we grant uncritically what has just been said, nuclear power should still be avoided. This since a significant, albeit small, likelihood of catastrophic events can never be excluded. First, the technology of even the most carefully constructed nuclear power plant is compatible with a remaining probability of malfunction resulting in a serious accident. Second, the so-called human factor (including everything from misconceptions in the initial design to mistakes in the daily management) constitutes a risk-source of unknown magnitude. To this must be added the waste problem, pointing out that although this can be dealt with in various ways, serious uncertainties regarding long-term consequences will always remain. In all these cases, likelihoods of hazardous events may be judged as very small. And even if we disregard the fact that there is also serious uncertainty regarding the actual size of some of the likelihoods, it may still be argued that the scenarios being risked through the use of nuclear power are so extremely undesirable that, in all, producing such risks is in fact highly irresponsible. This, I take it, is a fair summary of the basic standard argument made by the opponents of nuclear power.

---

[24]The development of so-called Generation IV nuclear reactors still being mainly at the drawing board stage, it is also unclear whether these sort of possibilities will bring new risk factors in the actual running of plants.

[25]I am here ignoring the doubts about the economic efficiency of nuclear power based on its potential insurance premium costs (which are astronomical and in most cases have been non-transparent due to them being carried primarily not by the industry, but by the nations housing the power plants). As will be seen, this benefit of the doubt will not prevent critical assessment of nuclear power from a precautionary point of view.

Immediately, we may note that this brand of precautionary reasoning runs the danger of neglecting some of the important aspects of the morality of risk imposition presented in the preceding chapter. As stated, the argument seems to presuppose the idea of forbidden risks – the notion of some scenarios being so extremely undesirable that it is always irresponsible to risk them, no matter the costs of preventing them and regardless of what risks would be produced by alternative options. However, first, it is far from obvious that even the most serious nuclear accident may actually qualify as such a scenario. Second and more importantly, the very idea of forbidden risks has been argued in preceding chapters to be seriously flawed in several respects. In order to have a sound ethical argument, we must be able to point to some alternative way of producing energy that would not bring risks comparable in terms of seriousness with the risks of nuclear power. Moreover, these alternatives must be so cheap and/or efficient that the level of well-being sustainable through the energy so produced is high enough for an acceptable mix of risks and chances being the end-result when all relevant factors are taken into account. For, otherwise, it is far from clear that our theory of the morality of imposing risks would in fact judge nuclear power to be an irresponsible source of energy.

This, of course, complicates the situation for the anti nuclear power argument. However, the difficulties do not seem to be insurmountable. If appropriately utilised, energy sources such as the sun, the earth, the sea, the wind and, of course, rivers and streams, may be claimed to be sufficient for upholding an acceptable level of well-being at the same time as the ways of producing energy from these sources do not produce any significant risks to environment or health (at least not in comparison to nuclear power). In all, these two aspects appear to support the claim that using nuclear power is a less responsible option than the use of these alternative sources of energy. True, it may be that if we use such energy sources rather than nuclear power we will be forced to accept a significantly shorter life expectancy and lower quality of life than what we are enjoying today (or, alternatively, that much fewer people are able to enjoy these goods). However, as long as a decent mix of risks and chances is secured, this may very well be an acceptable price to pay in order to avoid the risks of nuclear power.

Another line of argument that might be pursued on the basis of the theory I have been proposing connects partly to the one just spelled out, but focus especially on the possibilities of developing wholly new energy technologies. The argument above presupposes some of this, but it may be argued that this line of thought can be pushed further to support the idea of using significant resources for developing energy technologies that lack the drawbacks of nuclear power or fossil fuels, but that are capable of cost-effectively producing much more energy than technologies utilising the renewable sources mentioned. The typical example of this idea would be nuclear fusion technology, which is today researched and experimented with, but very far from being practically useful as an energy source.[26] It must then be

[26]See, e.g., the overview in Stork (2009). It should be underlined, that temporary enthusiastic bursts of individual scientists or groups (for a recent example, see Palmer (2010)) regarding the prospect

taken into account that, while a successful fusion technology would perhaps be able to resolve many dilemmas and problems highlighted above, it is highly unclear to what extent it is likely that any such technology will ever see the light of day even if vast amounts of money and time are spent on such a project. From the point of view of precaution it thus seems plausible to suggest that any large-scale involvement in the research and development of fusion technology would have to be preceded by a long stretch of preliminary investigations that are able to show the fusion idea to be feasible enough to warrant the costs involved. Thus, the wet dream of replacing the energy technologies of today with something that is much more efficient, cleaner and cheaper does not imply that we can dodge the problems we are facing today. These will most likely remain for a very long time ahead.

However, while apparently valid for highly developed countries, the lines of reasoning above regarding nuclear power are not obviously applicable to the poor countries of the world. For, once again, we may point to the apparent fact that the people of these countries suffer such a lack of wealth that the use of nuclear power for boosting their level of well-being would in fact be responsible in spite of the risks that are thereby created. So, while highly developed countries should abstain from the use of nuclear power and instead use alternative and less environmentally detrimental sources of energy, poor countries may indeed use nuclear power in order to increase their own level of welfare and thereby decrease the global inequalities in risk production. In order to work, this line of reasoning must assume that the use of the above-mentioned alternative energy sources would not suffice to secure an acceptable mix of risks and chances for the people of developing countries. Now, in light of the tremendous amounts of destitution in these countries, this claim may seem to be if not obviously true so at least very plausible.

However, even if the truth of this supposition is granted, there is a further problem. For, due to the fact that these countries are underdeveloped, they do not have access to the sophisticated technology, well-educated manpower and similar resources needed for upholding the same level of safety of nuclear power plants as a developed country. Therefore, we cannot make a simple comparison between the risks of nuclear power in highly developed countries and the risks of nuclear power in underdeveloped countries. In the latter case, the risks of nuclear power will be several times more serious, since a disaster will be much more likely and since the level of preparedness for handling the effects of such a disaster – not to speak of the waste management problem – will be much lower. In light of this, it is far from obvious that the use of nuclear power actually *would* secure an acceptable mix of risks and chances in such countries (although it may be true that this would boost the level of wealth and even out global differences in risk production).

---

of fusion technology has been a repeated pattern in the marketing of this domain of science, just as in the case of gene therapy (to be discussed in the next subsection), for several decades. Awaiting solid evidence, all such forecasts are to be taken with a substantial grain of salt. Something similar holds regarding the recent hype of the possible practical applications of synthetic biology (the feasibility of which very much remains to be demonstrated) – see, e.g., *Synthetic Biology & Bioenergy*.

In response to this, it may be suggested that developed countries should aid the underdeveloped ones, so that they are able to uphold a level of safety comparable to that of the former countries. However, in practice, this would mean that substantial amounts of wealth are transferred to underdeveloped countries and, if so would happen, it may be doubted that they could really be classified as underdeveloped anymore. In effect, it is doubtful if we would then be able to presuppose the low level of well-being needed for the argument in favour of the responsibility of using nuclear power stated above. This, then, leaves us with the crucial question of whether or not the fact that rich countries are overproducing risks at the expense of poor countries may justify that the serious risks of nuclear power are made substantially more serious on the assumption that this would even out the global inequality of risk production.

According to the theory of the morality of risk impositions defended in this book, an affirmative answer to this question is not ruled out in principle. However, in practice, whether or not such an affirmative answer would be forthcoming depends on complicated factors that may be assessed only through careful consideration of the specifics of each individual case, taking into account the relation of each such case to the global production of risks and chances as well as the effects of alternative strategies. And, of course, making such assessments will have a substantial price, since the destitution of poor countries will be sustained throughout the period when the investigations are undertaken – a fact making it very unclear whether or not such a level of precaution would indeed be worth its price and thereby responsible from a moral point of view. What is more, even if such investigations were to be undertaken, it is unclear whether or not the theory of the morality of risk impositions would be able to supply us with a unequivocal answer. This due to the indeterminacies built into this theory, which were explained in the end of the preceding chapter. I will return to the question of how to deal with the practical problems effected by this possibility in the next main section below.

Now, in addition to the more principled problems explained so far, there is also a problem of how to assess the use of nuclear power from the point of view of precaution that is of a more pragmatic nature. For, much as we may press the point that developed countries should indeed abolish the use of nuclear power to the benefit of non-polluting alternative sources of energy, this is not what actually happens in those developed countries where nuclear power is actually abolished (or not embarked on in the first place). On the contrary, in such countries, the preferred solution is to use oil, coal or gas – i.e., sources of energy that all contribute heavily to the pollution of the atmosphere. In countries such as my own – Sweden – this fact has come to be utilised as an influential argument in favour of the use of nuclear power.

There are some apparent similarities between this situation and the one regarding pollution trade discussed earlier. In both cases, there are strong reasons for a particular course of action (abolishing nuclear power and having a system for the limitation and redistribution of green-house gas emissions respectively). In both cases, this course of action may be undertaken in a way that is optimal (use non-polluting sources of energy instead of nuclear power and endow pollution rights according to the principle 'he who has not should receive' respectively) and other ways that are suboptimal. In both instances, finally, it is actually the case that when the desired course of action is taken, this is not in fact done in the optimal way. However, there is also one important difference. For while, in the case of pollution trade, it may be argued (by the pragmatist) that the suboptimal course of action may at least be a step on the way to the optimal one, this is hardly plausible in the case of nuclear power. Using oil, coal and gas as sources of energy can hardly be seen as a stepping stone to a forthcoming use of the sun, water, wind and the earth for this purpose.[27]

This last difference also makes for a difference as to how the problem should be analysed from the point of view of precaution. For while, in the case of pollution, a case may be made for the suboptimal solution on pragmatic as well as long-term strategic grounds, in the case of nuclear power, all that remains is the pragmatic conjecture. We must, according to this claim, surrender to the realities and accept the fact that living with the risks of nuclear power, although not ideal from the point of view of precaution, is still preferable to having the certain environmental destruction brought by the use of fossil fuels. However, this line of reasoning may be questioned in two ways.

First, it is an open question whether or not the risks of nuclear power are in fact preferable to the risks of coal, oil and gas. In order to assess this, a much closer examination of the involved risks and chances would have to be undertaken. Second, and more importantly, the pragmatic conjecture itself seems to me to be fundamentally flawed as an *argument*. Suppose that someone said to you: 'I could murder this person, but if I do not murder him I will in fact torture him for the rest of his life and, since lifelong torture is much worse than instant death, I am therefore justified in murdering him!'. Surely, your reaction would be that the person speaking has missed out on the essential point that he could avoid doing both! In much the same way, the fact that developed countries that avoid nuclear power in fact choose to use fossil fuels instead is no reason for using nuclear power even if it was to be proved that this source of energy is superior from the point of view of precaution to the other ones actually used. For, once again, they could instead choose to avoid both types of energy sources and instead opt for the environmentally less degrading ones.

6.2.3 Biotechnology

The field of biotechnology has expanded and evolved tremendously during the last four decades and there seem to be no reason to doubt the prospect for a continuation

[27]Unless we acknowledge a line of reasoning central to certain branches of classic revolutionary left-wing political theory; that boosting the flaws of an allegedly bad system may speed up the process of having it abolished. In the present case: making climate change problems much more serious much sooner than necessary may be a good thing in that such a development may motivate populations and political leaders to take effective measures against having the destructive process proceed even further. I trust that pointing out the flaws of this argument would be to overstate the obvious.

of this route of development in the future to be glimmering (as evidenced by the recent advances in synthetic biology).[28] While incorporating many areas of science and technology, as well as fields of application, the most important aspect of biotechnology is undoubtedly its potential for enabling the intentional creation, design and transformation of living organisms. It is this aspect that has delivered the example of genetic modification used to illustrate the basic issue of the price of precaution throughout this book. This example draws on what is presently the by far most advanced actual application of biotechnology – the controlled genetic modification of organisms such as plants, micro-organisms and animals in certain well-defined respects. However, in the future lingers even more advanced possibilities such as the elaborate *design* of organisms (i.e., not only modification of isolated functions, but of the totality of functions of an organism) out of elements found in nature and the actual *creation* of completely novel types of life constructed out of synthetically produced biological building blocks not present in nature before.

Since the current possibilities of genetic modification have already been illustrated by the example of genetically modified crop, I will not say much more about this. However, one additional, albeit important, remark is called for – actualised by the fact that most actual applications of these types of possibilities do in fact not imply the prospect of substantial benefits assumed in my example. Rather, these applications are aimed at creating commercially attractive products, thus stimulating the emergence and expansion of markets for the biotechnological industry. The typical example of this would be crops that have been modified to resist a certain type of pesticide, sold in conjunction with the crop by the same company, such as the "roundup ready" crops produced by the biotech conglomerate Monsanto. Other examples include so-called designer food, made to have various properties such as additional nutrition agents like vitamins or simply a more aesthetically pleasing look, taste or smell thought to attract the interest of consumers. Of course, from a technical point of view, other more desirable applications (like the one described in my example) are quite possible to imagine. On this note, it is not uncommon to hear biotechnological researchers describe visionary prophecies of how their knowledge and skill could contribute to the ending of world hunger and other equally praiseworthy objectives.[29] And, as mentioned earlier, the recent breakthroughs in

---

[28]Gibson et al. (2010) and Pennisi (2010). Another recent development in this trend of 'precision micro technology' is so-called *nanotechnology* – a field where the aim is to precisely control, manipulate and design extremely small entities (such as molecules, atoms or even sub-atomic particles). Although already applied in certain processes for the construction of materials, nanotechnology is still even more in the experimental stage than biotechnology. However, in the future, this line of development will most likely offer an even wider span of possible applications than current biotechnology and with these applications will, of course, follow new risks of various kinds. I believe, however, that much of what is said in this section about biotechnology will be applicable to nanotechnological risks as well. See, for example, Hunt and Mehta (2006), and O'Mathúna (2009) for introductions to nanotechnology, its possible risks and benefits, and the ethical considerations actualised by it.

[29]One example of such an idea that has already been put into some practice is the so-called Golden Rice. This genetically modified type of rice is an example of designer food that may have a more

synthetic biology have spurred even more ambitious scenarios.[30] However, what reality teaches us is that the drive to work in such directions is not very strong, due to weak economic potentials combined with the fact that research and development in this area require substantial financial investment over a long period of time.[31]

From a precautionary point of view, what this means is that for most biotechnological applications there is not much left of the dilemma-type situation created in my imaginary example of genetically modified crop. For while the long-term risks and uncertainties of large scale introductions of genetically modified organisms in the natural environment remain,[32] the price of precautionary measures that would force us to abstain from using the type of products we are presently considering is marginal. This for the simple reason that these products are just like any other commodity introduced on the market of a consumerist society. While this introduction (if handled skilfully) may indeed create an interest among consumers, the vast majority of these people (if not all) could have enjoyed just as much quality and length of life had this introduction never taken place.

Now, in order not to appear overly pessimistic, I must underline that if the uses of biotechnological products would actually serve any truly important objective, the situation immediately becomes more complicated. This is what is illustrated by my example of the genetically modified crop that could eradicate the environmental problems of farming. In that case, I have argued, a sensible solution, which is also supported by my theory of the morality of imposing risks, is that some time and effort should indeed be put into the project of clarifying the long-term risks and uncertainties. But I have also stressed that there is a limit to this recommendation: the time and effort invested must not be so long and great that the price of the precautionary measures becomes unacceptably high. It should be noted that should any of the great prophecies of biotechnological researchers, say the possibility of putting an end to world hunger or of supplying an everlasting source of non-polluting energy, become feasible, it could plausibly be argued that a sound morality of precaution would advise us to be even less resistant to start actualising such possibilities sooner rather than later.

---

far-reaching potential than merely attract the interest of already affluent consumers. However, its potential for solving important problems such as malnutrition is debatable. See, e.g., Mayer (2005).

[30] *Synthetic Biology & Bioenergy*.

[31] Alternatively (as regards, e.g., Craig Venter's vision of a limitless source of non-polluting fuel through synthetic microorganisms, see *Synthetic Biology & Bioenergy*), the economic potential of such developments, and thus attractiveness for investors, would require far-reaching patenting and commercial secrecy surrounding the work – factors that in turn tend to slow down, or even impede, the realisation of the applications (since scientific development works best when data and methods can be shared freely) and increase risks (since societal oversight and control becomes more difficult).

[32] In fact, if we consider the whole potential market for genetically modified products, the risks and uncertainties will plausibly be many times greater and this for two reasons. First, more risk-factors simply means that risks are accumulated into a higher total risk-level. Second, the more risk-factors that are present, that higher the risk for synergetic interactions between these that elevate the risks even more, or introduce additional risks of their own.

The visions of using biotechnology for alleviating environmental damage or world hunger are indeed appealing. However, I will now instead turn to an area of application for biotechnology where the prospect of significantly improving human conditions seems more realistic. This is the idea of biotechnologically changing not the human environment, but rather humanity itself.

With the exception of a few clinical experiments, none of the technologies described above have as yet been successfully applied to humans.[33] However, as the skills and knowledge of experts in biotechnology increase, we may safely assume that sooner or later the day will be here when the idea of biotechnologically transforming, designing or creating human beings starts to look as a feasible and appealing prospect. In connection to such scenarios, suggestions have been made to the effect that precautionary reasons support some type of limit regarding what is defensible to achieve with such means. In particular, it has been suggested that changes of the human genome should not proceed so far as to eradicate the human species and replace it with some other (perhaps in some respects improved) type of beings.[34] Obviously, ideas of this type could be underpinned by the forbidden risk approach, especially the version of this idea defended by Jonas, or a rigidity approach to the weight of evil inspired by a Jonas type concern for the continued existence of humanity. However, since both these suggestions regarding the morality of risk impositions have been rejected above in favour of the gradualist relative progressiveness approach, it is less clear how this type of suggestion should be assessed.

The central question to address on the theoretical basis accepted in this book regards the following: First, what benefits could be secured by the application of biotechnological transformation, design or creation to human beings? Second, what risks would be imposed by such applications? Third, how are these scenarios related to the idea of an acceptable mix of risks and chances?

To begin with, it is worth noting that most benefits that could be envisioned as the result of the type of biotechnological interventions under consideration would actually not effect the eradication of humanity. This for the simple fact that the vast majority of human characteristics that could be changed by such methods have no bearing on species membership. This, in turn, follows from the fact that, according to Ernst Mayr's classic definition, which is still widely applied to organisms procreating themselves through sexual reproduction, species membership is determined solely by the *reproductive potential* of such beings.[35] Hence, a being belongs to the human species if, and only if, it is the type of being that would under normal

---

[33] For a recent overview accessible to the lay person, see Häyri (2010, chapter 8). One strand of biotechnology that has been so applied for a rather long time, though, is techniques enabling the extraction of elementary biological (primarily genetic) information about humans. Applications of this kind, such as presymptomatic genetic testing and prenatal diagnosis, create quite a lot of ethical issues in their own right. See, for example Juth (2005), as well as Munthe (1996b, 1999c and 2007).

[34] Tännsjö (1990, 1993).

[35] Mayr (1942).

circumstances have been capable of producing fertile offspring through sexual interaction with other members of this species. A new species sees the light of day only if beings appear that lack this reproductive potential while they possess the potential for reproduction through sexual interaction with each other. This fact has two important implications: First, just as many benefits could be secured through the application of biotechnological interventions to humans without affecting their humanity as such, most of the risks that are imposed by such applications have nothing to do with eradicating the human species. Second, even if applications where to be made that would indeed exchange humanity for a new species, this need not impose very grave risks to life, limb, culture or the social order, since all that is needed for this upshot to be reached are minor changes of the reproductive mechanisms of humans. In conclusion, the scenarios that seem most important to consider from the point of view of precaution are those where humans are subject to vast changes by biotechnological means, regardless of if these changes affect species membership or not.

True, at the initial stage, even very small alterations of the human genome will be quite risky for the simple fact that there is a radical lack of knowledge regarding the reliability of applied procedures, as well as the interaction of genes, other biological materials, the surrounding environment and the ultimate effects of such processes on the resulting organism. These risks involve not only possible negative effects on the individual intervened upon, but also effects showing themselves in later generations through the inheritance of manipulated genetic components by the offspring of such an individual. However, on the view of the morality of precaution adopted in this book, these risks have to be balanced against the possible benefits of such interventions and how this balance relates to the idea of an acceptable mix of risks and chances. On this basis, it seems possible to argue that if the intervention has a real potential of alleviating a condition that affects well-being very negatively and where no alternative and less risky method for reaching such a result exists, taking the risks may very well be a morally responsible thing to do. Examples of such conditions could perhaps be the most serious genetic diseases, such as Tay-Sachs disease, where the affected person suffers from incurable anomalies in basic metabolic mechanisms that secure a very short life filled with extreme pain. However, it is equally obvious that this reason will weaken if we consider less serious conditions, partly since the risks run may then very well involve more grim prospects than those brought by the disease itself. If the seriousness of the condition is moderate rather than strong, we may also add doubts regarding if the risks are worth taking from the perspective of how far off from an acceptable mix of risks and chances a life with the condition will reside. Nevertheless, it may still be thought that precaution here requires us to strive for more knowledge, so that these less serious conditions could be approached by less risky means. And if this endeavour is successful, we may very well reach the point where biotechnological interventions become defensible from the point of view of precaution.

Now, these types of cases are a little bit like the example of genetically modified crop used earlier in this book, since the risky technology has the potential of alleviating serious risks. However, just as in the agricultural area, biotechnological interventions on humans could be undertaken also for other purposes.

In discussions of the ethics of genetically modifying human beings it is a popular move to attempt to distinguish between, on the one hand, gene therapy and, on the other, genetic *enhancement*. The latter notion is meant to capture such modifications of humans that would not alleviate health problems or other serious burdens, but that would nevertheless improve the human condition in various respects. The received opinion among many people seems to be that this distinction captures something of moral importance: While gene therapy may be defensible on the basis of a concern for human well-being, genetic enhancement is much more difficult to justify from a moral point of view.[36] However, several moral philosophers have demonstrated that there are in fact serious problems involved in upholding this distinction as a morally important one.[37] No matter what features we take to determine human success, flourishing or well-being, we can always argue that if the addition of such features in the case where someone suffers a loss of them is of value, surely, such an addition must also be of value in cases where there is no initial loss. In consequence, just as we should value the reparation of damages to the body or mind through genetic modification, we should also value the addition of psychological or physiological features that would improve human conditions beyond what we presently consider to be required in order to be of good health.

What this argument lays bare is the implicit assumption built into the distinction between therapy and enhancement that it is of significant value to improve the length and quality of a human life only up to a certain point. If some physical or psychological feature makes a person have a length and quality of life below this point it is considered to be a disease or disorder, the reparation of which would be a case of therapy that deserves to be valued highly. However if a person enjoys a length and quality of life on or above this point, changing any such feature in order to boost even further the length and quality of life of this person is not deserving of equal praise. But – and this is the thrust of the argument above – none of the suggestions regarding what constitutes improvements or impairments of the value of life support the idea of such a point, let alone hints at any way of justifying suggestions regarding where such a point is to be located.

This problem has led some more recent commentators to pursue the idea that there should be no limits in terms of the seriousness or nature of conditions with regard to the application of gene technology to human beings. As long as they are voluntary undertaken and reasonably safe, such applications are morally defensible and should be allowed. In some cases, the argument that there is no point demarcating therapeutic applications of genetic modification from enhancing such applications has even been held out as a wholesale defence of any type of human genetic modification.[38] However, as will now be seen, such conclusions appear quite

---

[36]See the brief but excellent overview, description and further references given in Malmqvist (2008, especially pp. 26–27).

[37]See, for example, Glover (1984), Harris (1998), Munthe (2000), and Tännsjö (1993).

[38]See, e.g., Agar (2004) and Boström and Roache (2007). In addition, these debaters – as well as somewhat more cautious ones (e.g., Buchanan et al. 2002) – tend to project a surprisingly unreflected and/or simplified idea of what may actually constitute an enhancement. There are

frivolous once we apply a defensible ethics of risk to the condition of safety just mentioned.

Applying the idea of the morality of risk impositions defended above, it transpires that this idea in fact supports the idea of a morally relevant demarcation that lacks the feature of insisting on a morally relevant basic distinction between therapy and enhancement, but nevertheless reminds of this idea in its applications. In virtue of the relative progressiveness idea and its inherent reference to the notion of an acceptable or decent mix of risks and chances it becomes possible to say that there is a morally significant difference between genetic modifications that would improve the conditions of a person so as to make her enjoy an acceptable or decent mix of risks and chances and such modifications that would improve a person's conditions beyond this point. And this difference holds even if we assume the improvement to add to the respective persons' lives equal amounts of longevity and/or quality. Thus, if we assume the risks involved in these two procedures of genetic modification to be on a par, taking such risks in the first case would be less irresponsible than in the latter case. In consequence, even if a particular genetic procedure that indeed has a substantial chance of improving a person's length and/or quality of life closer to an acceptable or decent mix of risks and chances could be justified from the point of view of precaution, it does not follow that genetic procedures that would promise an improvement of an equal degree but *beyond* such a mix could be so justified.

Of course, if the risk scenarios involved are different, the outcome of applying my theory of the morality of imposing risks will also be different. Suppose, for example, that the genetic enhancement procedure is much less risky that the therapeutic procedure. In that case, the difference as to the degree of responsibility involved in applying these methods would, of course, diminish. However, we can still retain the basic idea of a morally significant distinction between what may often be referred to as therapy and enhancement with reference to the relative progressiveness idea and the notion of an acceptable mix of risks and chances. For even if the enhancement procedure in question would not be very risky, it still holds that if equally risky procedures could instead be applied in order to bring those people residing below an acceptable mix of risks and chances closer to this point, choosing the latter activity would be the more responsible choice to make.

One important lesson of this application exercise is that it uncovers a basic flaw in the conceptual prerequisites of the ethical debate about gene technology that is formulated in terms of a therapy-enhancement distinction. Supporters and critics alike seem to assume that either there is an absolute ethical line to be had in terms of such a distinction, or there is no ethical line to be had at all. This assumption, however, has just been demonstrated to be implausible. In particular, the strategy of reductively framing debates on the ethics of gene technology as having to assume a simplistic dichotomy of possible positions presenting an allegedly unavoidable choice

---

many examples given of both physical (such as strength and endurance) and cognitive/intellectual abilities (such as memory or computational skill) with no necessary or obvious connection to what makes life better for a person and the boosting of which may in fact make life worse. For elaboration of this type of point, see Shickle (2000).

between "bioconservatism" and "transhumanism"[39] can henceforth be viewed as an elementary misconception.

Before closing this discussion of genetic modification of humans, three important features of the points made should be noted. First, it is not certain that a distinction between different genetic modification procedures along the lines suggested above will always correspond to the intuitive divisions of genetic procedures in terms of therapy and enhancement. In particular, taking into account what serious health problems are in fact burdening most of the people on our planet, it may very well be the case that some of the genetic conditions perceived as health problems in developed countries do in fact not rob people of an acceptable mix of risks and chances. In effect, what would be seen as gene 'therapy' in a typically Western setting may rather be classified as a case of 'enhancement' when analysed from the point of view of precaution on a global scale. This, in turn, may be due to two different facts.[40] On the one hand, we may have an idea of what is to count as an acceptable mix of risks and chances that is non-relative and, on this idea, some of the hardships considered as health problems in a Western setting do in themselves not impede people from enjoying an acceptable mix of risks and chances. On the other, we may entertain the idea that what is an acceptable mix of risks and chances is relative to what is at stake in a particular situation of choice. So, for example, if we have the choice between spending our resources on applying risky methods of gene therapy for the eradication of a genetic mutation that increases the risk for mild depression and giving those millions of people in need of that access to not very risky, efficient protection against and treatment of malaria, it would seem that the morality of precaution urges us to go for the latter.

Second, what has just been said also demonstrates that if we go for the relativistic solution to the problem of how to determine the acceptable mix of risks and chances, the distinction advocated above is not fixed in its application to various conditions. Rather, it is highly dynamic in its view of what is to count as responsible or irresponsible depending on what are the actual options in and circumstances surrounding particular situations of choice. Suppose, for example, that the horrors of malaria had already been mastered. In that case, obviously, the second option in the just described example would be a sad waste of resources and if the only alternative is to go for the option of gene therapy for susceptibility to mild depression, it would suddenly look as this could be seen as quite responsible. This is, of course, not to say that such a measure would be defensible from a precautionary point of view (the risks and/or uncertainties may still be too severe), but the analysis of this question would have to be made on the assumption that susceptibility to mild depression indeed does rob people of a decent mix of risks and chances. This, of course, is an important difference to the idea that such a mix is determined non-relatively.[41]

---

[39]Boström and Roache (2007).

[40]Cf. the considerations discussed at the end of Chapter 5.

[41]In Chapter 5, I suggested that, perhaps, relative and non-relative determinants of an acceptable or decent mix of risks and chances should be balanced against each other in some way. Obviously,

Third, what is further illustrated by this is that the result of applying the relative progressiveness idea in order to justify a moral distinction, that may substitute the idea of distinguishing enhancement from therapy, is as unclear as the notion of an acceptable or decent mix of risks and chances itself. Besides the issue of relative or non-relative versions of this notion, we also have the basic issue, mentioned in Chapter 5, of how many possible outcomes that should be seen as securing such a mix. As mentioned in that context, this issue ultimately would have to be settled on the basis of more general considerations regarding the morally proper degree and price of precaution.

6.3 Policy

One important conclusion reached in the first section of this chapter was that the ideal of precaution defended in this book strongly calls for an outright political (rather than, e.g., purely educational) set-up in order to realise the requirements of this ideal. However, the discussion of the problems of various activities, technologies and social practices undertaken in that and the preceding section has not in itself answered the question of how such a *policy* of precaution should be designed. While various points have been made about different problematic areas and in what direction a morality of precaution seems to suggest the avoidance of these problems to reside, nothing has as yet been said in detail regarding what more exact political instruments would seem adequate in order to move things in the desirable direction.

In the present section, I will start investigating this issue within the framework of a nation. This investigation will actualise a number of questions and problems highlighted in earlier sections and chapters. By revisiting these, a partial solution to the problem of what a precautionary policy should look like in a national setting will be sketched. This sketch will, in turn, point to a number of general policy issues that go beyond the framework of isolated nation states. These issues will be dealt with in the next main section, together with various other general issues regarding the more far-reaching practical implications of the ideal of precaution defended in this book.

6.3.1 Do We Really Need a PP?

As mentioned in Chapter 1, several commentators in the debate on PP has expressed serious doubt as to the meaningfulness of the notion of *a* PP. Per Sandin, whose inquiries into the content of PP have been referred to in various parts of my own investigation, has reached the conclusion that the prospect of finding an 'authoritative formulation' of PP seems meagre to say the least. Similarly, Marko Ahteensuu

---

such an idea would further complicate the application of the relative progressiveness idea to the cases at hand.

more recently claims that, at best, separate very specific and concrete PP's designed for certain particular areas of activity (such as agricultural use of genetic modification) might be possible to justify on ethical or political philosophical grounds in combination with factual considerations.[42]

In one sense, these conjectures would seem to run contrary to the claims that I have been defending in this book. However, also my own theory of the morality of precaution indeed seems to support these sceptical claims to some extent. On the basis of the complexity of the issue of precaution that has gradually revealed itself through the construction of the theory, as well as the applications of this theory undertaken earlier in this chapter, it seems plausible to discount any hope of finding one single principle that could function as a politically useful practical decision tool capable of meeting the requirements of the ideal of precaution that I have been defending in any imaginable social setting and situation of choice. However, this is actually no reason for despair. Or, so I will now argue.

In order to see why this is so, it is useful to consider why one might take the just mentioned scepticism to *be* a reason for despair. I take it that there are two lines of thought pulling in this direction. The first of these is the claim that if there is no single version of PP to be had that is both justified and sufficiently intelligible for practical purposes, then the very idea of an ideal of precaution spelt out in Chapter 1 of this book has proven itself incoherent or, at least, implausible. This, however, ignores the basic point, made already in Chapter 1 and further developed in Chapter 2, that an ideal of precaution taken to be expressed by the notion of PP had better be seen as separate from PP itself. Rather, this ideal should be seen as an underlying norm spelling out requirements to be met by any plausible version of PP. In this light, the fact that several different and mutually incompatible PP's may be so justified in different circumstances in no way threatens the notion of a plausible and coherent moral ideal of precaution, implied by a plausible theory of the morality of imposing risks.[43]

The other line of thought is the idea that, even if the just said would be true, such an ideal of precaution could never be used for any practical purpose unless it could function as the basis for a PP (or several separate PP's specifically designed for particular areas of practice) capable of filling the role of a politically feasible vehicle of practical implementation. If that is so, the conclusion that what may at best be justified is a cluster of different generic PP's is bad news indeed, since principles employed in practical politics and policy making for various reasons need to be simple and clear-cut. However, even if that was to be true, this line of thought rests on the crucial assumption that a practical vehicle for the political implementation of the ideal of precaution has to take the form of one or several principles. This, in turn, rests on a presumption that seems to be shared by many participants in the debate on PP – a presumption that I will now proceed to question.

---

[42] Sandin (2004, p. 24), and Ahteensuu (2008).

[43] This conclusion seems to be in line with Ahteensuu's point (2008).

6.3.2 Principlism vs. Proceduralism

A very clear expression of this presumption is brought out in a paper by Sven-Ove Hansson and Per Sandin, where they specify a number of requirements that they claim have to be met in order for a practical application of PP to be successful.[44] The basic model there described is the idea of formulating PP as a policy-guiding decision rule of the form "if the facts F obtain, then decision D should be made". Practical application of PP in this model is achieved by specifying for any particular case the variable F and then deducing a specific description of the variable D for each such case. For this type of application to be successful, Sandin and Hansson observe, several requirements have to be met. For example, it must be possible to deduce a specific description of D in every particular case and this description must be specific enough for one decision in particular to be prescribed by the application. I will henceforth call this model of practical application of PP *principlism*.[45]

If principlism is taken for granted it seems clear that the results of my investigations do provide reason for despair. This since the basic theory of the morality of precaution worked out in the preceding chapter is vague in several respects. This, in turn, implies that a decision rule of the form propagated by Sandin and Hansson that could be supported on the basis of my theory may be expected to be vague too. In practice, this means that we have no reason to expect that a procedure of application of the principlist type will be able to produce unique and clear-cut prescriptions for any case that may arise in practical policy making. In some cases, no prescription at all may follow and, in other cases where such a prescription would follow, it will not recommend one unique decision in particular.

Since I have argued that a plausible theory of the morality of imposing risks will have to involve vagueness, this may seem to support the sceptical conclusion. There is no normatively plausible version of PP that is also practically useful and, consequently, the vision of formulating an ideal of precaution capable of addressing the practical needs spelled out at the outset of this book may be laid to rest on the scrap-heap of intellectual history. This, however, presupposes that principlism is the only model available for practical application of such an ideal of precaution. But, in fact, there are alternative ways of envisioning how a practical politics of precaution may be structured.

---

44 Sandin and Hansson (2002). Peterson (2006) seems to be assuming something similar when setting out his formal conditions for any satisfactory version of PP.

45 The choice of this term is inspired by the use of the same notion in debates on the methodology of medical ethics (see, e.g., Davis 1995). However, in that context, principlism is juxtaposed to casuistry and virtue based ethics (with regard to what basic normative outlook to adopt to medical ethical issues), rather than the use of practical policy procedures for solving some of the problems involved in the practical application of moral ideals.

Instead of principlism, we could opt for what I will call *proceduralism*.[46] While principlism approaches practical application from a perspective where the aim is to find theoretically mechanical solutions to practical problems on the basis of norms (in this case, PP), proceduralism instead approaches application as a truly *practical* endeavour that has to be *performed by people*. That is, application necessarily involves human beings making use of their capacity for judgement. To be true, since it is application we are talking about, there are theoretical considerations that are being applied by these people. However, proceduralism does not presume that these considerations can produce solutions to practical problems by themselves by way of simple logical inference. They might do so at times, but it may also be the case that the decision makers will have to use their capacities for judgment in order to reach a practical conclusion on the basis of the available theoretical considerations.

Seen from this perspective, the answer to the question if we really need a PP is negative. What is needed is plausible theoretical considerations that may guide decision makers also employing their own judgement in specific cases. We do not need a precautionary *principle*, we need a *policy* that expresses *a proper degree of precaution*.

In practice, then, determinations of whether or not some practice is defensible from the point of view of precaution may be structured in the following way: The theory of the morality of imposing risks provides theoretical considerations regarding what type of facts are relevant for such a determination and suggests how they should be balanced against one another in decision making. These considerations are used as guidelines by policy making bodies, the members of which are human beings who, in particular cases, also have to employ their own judgement. Anyone who has been involved in practical policy making will find this model familiar indeed, since this is how policy is actually made in reality.

I will not enter any discussion of how the procedural model should be designed in detail. Presumably, different designs may fit different policy areas. In some cases, the decisional body may be the highest institution of political authority, such as the government or parliament. However, since legislation usually is a rather inflexible tool of policy making and since it is important from a precautionary point of view to have a policy that allows for activities previously judged as irresponsible to be allowed on the basis of new and favourable evidence, this level would presumably mostly be involved at the initial stage of *setting up* (rather than managing) a precautionary policy. In most cases, therefore, these institutions may instead choose to delegate decisional authority to specialised units such as committees or subordinate authorities, which may be engaged in a continuous re-evaluation of activities falling within their jurisdiction. In yet other cases, the appropriate solution may instead be to transcend national borders and hand over decisional authority to multi-lateral institutions (I will return to this possibility below). The main point is that whatever

[46]Not to be conflated with Rawls' notion of procedural normative theories of justice (see Rawls 1971) or positions, e.g., in health care and research ethics inspired by this notion that have recently attracted criticism (see, e.g., Ashcroft 2008).

particular design is chosen, it is always possible for the considerations of the ideal of precaution defended in this book to guide decision making.

Before moving on to more specific issues of how policy making along these lines may use precautionary considerations, I will comment briefly on two features of the procedural model that may also be seen as drawbacks. First of all, it is obvious that this model will have to deal with the familiar phenomenon of conflicting opinions and judgements. If the personal judgement of policy makers is to do a part of the job of application, this means that we run the risk of ending up with decisional paralysis, since conflicting judgements may impede the production of a decision. Or, so it may seem.

However, decisional paralysis becomes a real problem only if it is systematic (see Chapter 2) – occasional cases of failure to reach a decision are acceptable, unless the circumstances are exceptional. Moreover, systematic decisional paralysis due to conflicts of judgement can be safeguarded against by choosing an appropriate political design. That is, what we learn from this objection is mainly that the decisional bodies that are to perform the actual procedures of decision making have to have access to tools that may be used for resolving the conflicts so that a decision is reached in most cases. This is what politics is all about: to find ways of peacefully resolving conflicts on practical matters. And the means available for attaining this are familiar enough: negotiation and, in the end, voting procedures.[47]

Secondly, actual precautionary policies according to the procedural model will partly depend on the individual judgements of policy makers and, as we have just seen, on the outcome of political power struggles. Since such policies will be a part of a politics of power, this may be seen as a problem from the point of view of legal security. The line of reasoning suggesting this proceeds something like this: Precautionary policies will be about the use of societal coercion and force against individual parties – if not complying with the policies, these parties may be subjected to punishment or other types of societal sanctions. But it is a basic requirement of any decent society that individual parties should be able to foresee what actions on their part would trigger such responses from society and that these responses should not be arbitrary. That is, legal security requires that no one is subjected to societal force unless there is a societal rule to this effect that enables everyone to identify what actions would fall under the rule in question.[48] But if the content of precautionary policies will depend so much on personal judgement and accidental political power balances, this requirement will not apply.

I agree that if this was true, it would indeed constitute a powerful argument. However, there are two reasons for rejecting this suggestion. First, there are reasons to expect that any precautionary policy structured according to procedural lines will eventually enable individual parties to foresee the content of the policy. This is due to the fact that we may expect the decisional bodies to gradually

---

[47] Seen from this perspective, one might say that the idea of principlism expresses the impossible dream of making political decisions without politics.

[48] See, e.g., Tännsjö (1999) for an account of legal security along these lines.

develop coherent practices regarding how to judge individual cases that share similar features.[49] Secondly, and more important, the argument presupposes a design where the decisional bodies shaping the precautionary policies function like courts of law – intervening only retrospectively when some potentially questionable action has been identified. However, it seems to be a lot more sensible and in tune with the general intuition of precaution to have the decisional bodies function as prospective givers of permission. That is, the basic tenet of the policy is that if you contemplate any activity that meets certain criteria, then you are prohibited from undertaking this activity unless you have been given specific permission by the body in question. So designed, the precautionary policy will easily comply with the requirement of legal security as long as it is sufficiently clear what types of activities are in need of such a permission. This way of handling things is familiar from many different areas of policy making where important interests are at stake, such as the licensing of pharmaceuticals, industrial chemicals, and so on.

6.3.3 De Minimis Revisited

The proceduralist solution to the practical political application of the ideal of precaution just spelt out necessitates that we again consider the idea of *de minimis* risk. As was demonstrated in Chapter 2, the question of how grave or serious the initially perceived risks of some activity must be in order for the other considerations of the requirement of precaution to 'kick in' is just one of several factors determining the price and degree of precaution of some policy built on this requirement. So the suggestion is not that the procedural solution to the problem of application makes this factor more *important* than the other considerations. However, since this solution will have to work from a set of initial criteria stating under what conditions some activity will need special permission (accorded or not accorded on the basis of judgements employing ideas about the moral responsibility of imposing risks), the question of how initially dangerous an activity must appear in order for permission to be required becomes practically prior.

There are four considerations from the earlier discussion that seem to have special relevance to this issue. First, in Chapter 2, I suggested that a plausible idea of *de minimis* risk should be based not only on considerations of the probability-aspect of risks. That is, whether or not the apparent risks of some activity should be seen as serious enough for extra measures aimed at preventing these risks to be required should depend on both the seriousness of the harm that would ensue, should the risk be actualised, and the likelihood of this actually happening. In my view, this

[49] This gives room for some legal insecurity until such coherent practices have been established. If the time-frame for this is not too long, it is hard to see this as a powerful objection, however. For example, compare with the case of agencies for the licensing of pharmaceuticals. When these were created around the world (in the 1960s) and given the power to decide which pharmaceuticals are legal and which are not, not much of coherent practice was in place, but with time such practice developed.

initial suggestion is supported by the general spirit of my theory of the morality of imposing risks. Especially so since this theory emphasises the moral importance of avoiding additional harmful events in virtue of the idea of relative progressiveness. In conclusion, therefore, the criteria employed in a practical policy of precaution for deciding when an activity needs special permission should involve considerations of both the likelihood of harmful events and the seriousness of harm that would ensue, should these events actually occur, with some priority given to the latter factor.

Secondly, one important basic consideration of the theory of the morality of precaution developed in this book is the idea of the value of improving the evidential basis of decisions. This notion has important implications in the present context since it supports the suggestion that if the initial assessment of the risks of some activity is based on weak evidence this is a reason for postponing the activity and collect further evidence. In effect, even if the potential risks of some activity do not look very serious there may still be good reasons for abstaining from the activity, namely if the evidential basis for this assessment is weak. This is an idea of the notion of *de minimis* that departs considerably from the standard use of this notion in practical decision making, where lack of evidence is regularly used for rejecting demands for extra caution.

Thirdly, still we must not lose perspective of the importance of considering the effects on the price of precaution. This means that the notion of *de minimis* must be employed also to the precautionary measures themselves. That is, an acceptable policy regarding *de minimis* must be moderated in view of the overall costs (in terms of actual harm, lost benefits as well as the introduction of risks thereof) of employing restrictive measures against potentially dangerous activities. It is not in tune with the general spirit of precaution if a precautionary policy itself produces more harm and creates more serious risks than it is able to prevent. In view of the analysis undertaken in Chapter 2, what this means in practice is that the choice of criteria of the *de minimis* type actually included in a precautionary policy has to be made on the basis of a delicate balancing of the inclusiveness of these criteria with regard to what activities would be subjected to regulative measures and what requirements these activities must meet in order to receive permission. The tougher these latter requirements, the more reason there is for applying less inclusive *de minimis* criteria and vice versa. Ideally, this balancing should be made so that the overall policy prescribes a proper degree and price of precaution in light of the theory of the morality of imposing risks sketched in Chapter 5.

Fourthly, what has just been said in turn implies that the inclusiveness of the *de minimis* criteria may justifiably vary considerable between nations and this in two ways. On the one hand, different nations may choose different mixes of the inclusiveness as regards potentially dangerous activities and the strength of the requirements that must be met by those activities that are so included. On the other hand, as we have seen earlier in this chapter, the relative progressiveness idea supports the suggestion that different nations may justifiably prescribe different degrees and prices of precaution, depending on the overall mixes of risks and chances available. So, given one set of requirements that have to be met by included activities, some nations may be justified in applying less inclusive *de minimis* criteria than others in order to be better able to secure or, at least, approach an acceptable or decent mix of risks and chances. However, nations that already have secured such a mix instead have reasons to employ more inclusive criteria and maybe also stricter requirements for the activities included in order for these to receive permission.

6.3.4 Justifying the Proof Requirement of Justifiable Policy Claim

On the basis of what has been said so far, I will now conclude this section on national policy by reconsidering some further ideas that have been touched upon earlier in this book and that have been given quite some attention in the debate on PP. The first of these ideas is the *proof requirement of justifiable policy claim*, formulated in Chapter 1 and, as we saw there, included in several policy statements aspiring to express the notion of PP, e.g., in the *Rio Declaration* and within the EU. This idea was formulated thus: Policy measures against some activity that may bring great harm may be justified even if there is no scientific proof that this activity imposes (or would impose) this harm.

It is rather easy to see that this idea can be justified on the basis of what has transpired in this and earlier chapters. Lack of scientific proof or evidence regarding riskiness is not the same as lack of riskiness. On the contrary, in many cases, what absence of evidence or proof demonstrates is that the evidential basis for making decisions is weak. Since we have reason to improve our evidential basis regarding risks and potential benefits when it is weak and when this can be done without effecting unacceptable costs, policy measures aimed at this end may be justified from the point of view of the morality of precaution.

However, it is important to note that this is quite compatible with actually allowing very risky activities. What the requirement of justifiable policy claim says is merely that restrictive policy measures *may* be justified, not that any such measures *are* justified. In order to reach the latter conclusion, the overall price of the precautionary measures has to be assessed and evaluated according to the view on the morality of imposing risks. So, in all, justifying this requirement does not take us very far on the road to an acceptable precautionary policy. In this book, I have made a string of suggestions aiming at supplementing this minimal idea with further requirements to this effect.

6.3.5 Justifying the Burden of Proof Requirement

The second idea to reconsider is the *burden of proof requirement*, which in Chapter 1 was formulated thus: Showing that some condition for the permissibility of activities is met is the responsibility of those who propose to undertake the activity in question. Also this idea is a recurring theme in actual regulations of many nations taken to express the notion of PP. However, it is less of a general principle underlying such regulations (as may be said of the *proof requirement of justifiable policy claim*) than an administrative rule employed by these regulations. As observed in Chapter 1, the *burden of proof requirement* may be applied to any regulative policy regardless of whether or not it is taken to be an expression of PP. However, in the following, my comments will only regard the context of precautionary policy making.

What this rule implies in the context of a precautionary policy along the lines suggested above is that it is not society that should be burdened by the obligation to demonstrate that an activity falling under the precautionary regulation should be permitted. Rather, for such activities, the principle "guilty until the contrary has been proven" applies.[50] In that way, the burden of proof requirement seems to follow swiftly from my proposal. However, one important exception should be noted.

The type of policy I have suggested contains two parts: First there is the part of inclusion, expressed by the *de minimis* criteria discussed above. Activities included under the policy area according to these criteria must meet the other requirements of the policy in order to be permitted. This is the second part of the policy. And it is not society that carries the burden of proof for demonstrating that the requirements of this second part are in fact met, but those who want to undertake the activity in question. However, regarding the first part of inclusion, society does carry the burden of proof. That is, if someone engages in an activity and society wants to prosecute this on the basis of this activity lacking appropriate permission according to the precautionary policy, it is society that needs to demonstrate that the activity in fact meets the criteria for inclusion.

The main reason for this limit on the *burden of proof requirement* is the considerations of legal security mentioned above. However, besides this, there also seem to be general precautionary reasons for such a policy design. A policy requiring that each and all of us would have to prove our innocence in terms of precaution whatever activity we engaged in would prescribe a much too high price of precaution in that it would seriously threaten to impede all forms of social progress – also those forms that would seem to be highly desirable from the point of view of precaution.

6.3.6 Conservatism Revisited

In Chapter 2, I considered what seems to be a rather common, albeit unreflected, conservative interpretation of PP: the idea that precautionary policies should only regard *new* activities. I rejected this idea of what I called simplistic conservatism on the basis that it gives *status quo* a much too privileged position in societal decision making. From a precautionary point of view, whether some activity has been undertaken for some time or if it is a novelty is in itself of no consequence for how it should be evaluated in terms of the morality of the risks it imposes and the benefits it may secure. This basic line of thought has been preserved in my theory of morally

[50] A reminder to the reader is in order: "proven" may be interpreted in a variety of ways. See further, Chapter 2.

responsible risk impositions. However, the policy design sketched above may seem to reintroduce a conservative element. Since the design is built on the idea of having some authority granting or not granting permission to proposed activities meeting the inclusion criteria, it may seem that the practical application of this precautionary policy will as a matter of fact favour *status quo*.

Ideally, this should in fact not be the case. The precautionary policy sketched so far applies also to activities that have been undertaken for some time. In other words, the authority or authorities implementing the policy should consider not only novel activities, but also such activities that are already a part of social reality. Ideally, then, the policies should include not only prospective but also retrospective and, possibly, revisionary action. However, I am aware that political realities will probably not be able to live up to this ideal, at least not immediately. There is a strong human tendency to resist changes of *status quo*, at least when *status quo* appears to be somewhat beneficial and this tendency seems to have a quite strong impact on political decision making. For this reason, we should expect that the actual application of precautionary policies along the lines suggested above will express more of conservative sentiments than what could actually be justified, at least initially. This calls for two comments.

First, it seems to me that awareness of this problem may be built into the precautionary policies themselves. It may be put into the statutes or instructions of regulative agencies that they are to evaluate not only new activities but also *in a piecemeal way* audit current ones. To some extent, the evaluation of new activities will have to include this (since whether or not the risk of an activity is responsible to impose or not partly depends on what risks would be imposed by alternative activities, including the alternative of preserving *status quo*), but the conservative tendency mentioned above provides a reason for stressing this in an explicit way. The policy will then in practice allow for some initial conservatism, but piece by piece the room for this tendency will shrink as *status quo* is gradually being assessed from the point of view of precaution. This, of course, presupposes a rather long-term perspective being applied in precautionary policy making, a point I will return to in the next section.

Secondly, it may be that the conservative tendency that will express itself in practical decision making need not be as unjustified as initial appearance may suggest.[51] One aspect of this has to do with lack of knowledge: in many cases we know considerably more about *status quo* than about various possible changes of it. Partly, this depends on the simple fact that we are less familiar with the latter, but it also seems to be the case that the uncertainties and gaps in our knowledge regarding the effects of some change of *status quo* expand dramatically the more far-reaching such changes we consider. Since basing risky decisions on weaker evidence is a cost according to my theory, this means that *status quo* often enjoys the privilege of not being burdened by such costs to the same extent as proposed changes. True, this privilege may often be balanced out by other considerations, for example, if our

[51] I owe Ingmar Persson for bringing my attention to this.

knowledge about *status quo* is that it is very harmful or risky. Moreover, it should be underlined that, in many situations of considerable relevance from a precautionary point of view, our feelings of being more familiar with *status quo* than with proposed novelties may in fact be severely misleading. This since such sentiments as a rule are based on experiences originating from circumstances that may very well no longer apply without us knowing anything about that. However, it still seems to hold that, unless there are special reasons, precaution will in practice tend to favour less dramatic changes of *status quo*.

Another aspect is the following: An important part of my theory of the morality of imposing risks is the relative progressiveness idea. It seems to be a consequence of this idea that in societies where people already enjoy an acceptable or decent mix of risks and chances, to introduce additional (unclear) risks is much more difficult to justify than in social settings where people are not in such a favourable position (assuming that the activity producing the risks will also produce some benefits). In the former type of societies, therefore, it seems that resisting changes of *status quo* is easier to justify from a precautionary point of view. This does certainly not mean that all risky activities in such societies should be allowed to continue – we may very well have very good reasons for improving the situation also in settings where an acceptable or decent mix of risks and chances is secured. However, given that all changes for the better will also produce additional risks, it will take more to justify any such change compared to settings where an acceptable mix is in fact not secured. Most important, however, in the more favoured type of societies, introducing new risky activities will be considerably more difficult to justify from a precautionary point of view.

Now, since the notion of an acceptable or decent mix of risks and chances has not been defined in any precise manner, it is difficult to tell what this line of reasoning implies for practice in more concrete terms than those presented so far. Moreover, even if we did have access to a precise definition, we would still have to sort through a mountain of complicated empirical questions. Nevertheless, even if my rejection of simplistic conservatism still stands, this revisitation of conservative ideas still shows that some elements of (reflective rather than simplistic) conservatism in practical policy making could indeed be justified.

6.4 Big Questions

Having considered a variety of different cases and the issue of policy making in the context of nations, I will now close this final chapter by addressing a number of more overarching and principal issues that have in various ways been actualised during the previous discussion. What unites these issues is that they concern the extent to which the adoption of a precautionary ideal necessitates some sort of 'paradigm shift' in our basic political ideals. Starting out by revisiting the enlightment ideal, I will then move on to the question of what basic values precautionary policies should operate on and the issue of the desirable level and time-horizon of precautionary policies (both of these issues where introduced in Chapter 2 and then left open throughout most of the discussion so far). Together with some further points that have transpired in the present chapter, the consideration of these issues, I will argue, makes a case for *cosmopolitan* precautionary policies – at least regarding policy areas with strong global aspects, such as many environmental issues. This conclusion calls for some comments with regard to the extent to which a precautionary policy can honour the political ideals of liberal democracy. I will argue that, while taking precaution seriously certainly poses a challenge to some aspects of contemporary liberal democracy, at the same time, the cosmopolitan model of precautionary policy making itself offers ways of preserving – in some aspects indeed strengthen – the most basic and important liberal and democratic values.

6.4.1 The Enlightment Ideals Revisited

In Chapters 1 and 3, I briefly touched on the issue of how the notion of PP and the underlying idea of a precautionary ideal relates to the ideals of the enlightment. As mentioned, it has been a recurring theme in discussions about PP that this notion in some way constitutes a radical break with the era of 'modernity' – that historical period marked out by being driven by the enlightment ideals. To some extent, this suggestion has indeed been shown to be warranted, but at the same time the precautionary ideal still preserves so much of the basic ideas of the enlightment that it had better be seen as an evolvement. Or, so I will now argue.

A driving force of the enlightment is the ideal of (instrumentally) rational and scientifically informed policy making in the light of whatever values are found worth pursuing through democratic processes. The basic idea of a precautionary ideal challenges this tradition by insisting on, first, that rational requirements on policy making should be supplemented by moral ones. Even if the basic goals of a policy are taken for granted, there are further moral features required by the precautionary ideal for this policy to be justifiable (i.e., the policy should not impose risks in a morally irresponsible way). Secondly, these additional moral requirements on policy making may very well mean that the (instrumentally) rational solution to a particular policy issue is not the one that should be chosen. Thirdly, the relation of science to policy making in the precautionary model becomes less straightforward in that lack of scientific evidence for the riskiness or harmfulness of some activity does not necessarily speak in favour of this activity. Fourthly, as will be further developed below, the enlightment idea of neutrality as to what goals should be pursued in society is to some extent at odds with the precautionary ideal, since this ideal presupposes a definite set of basic values to operate on.

In other aspects, however, the precautionary ideal is indeed able to honour many core elements of modernity. The moral requirements of precaution will in many ways have to make use of both the models of (instrumental) rationality and science. As regards the former, the gradualist approach to precaution defended in Chapter 5 retains the basic idea of calculated risk taking. Simply put, there are no upshots that we are forbidden to risk no matter the circumstances and there are no options that we are always permitted to choose no matter the circumstances. This means that the models developed within decision theory for balancing values and likelihoods of all available options will continue to be useful within a precautionary policy, albeit complemented by additional elements. Moreover, the idea of a moral reason to improve the evidential basis of decisions when this can be done at acceptable cost supplements the basic model of rational action with regard to an issue on which it has had a hard time indeed to provide a plausible solution.

In fact, this last element means that the precautionary ideal in fact *strengthens* the role of science in policy making. Precautionary policies will as a rule require more of scientific investigation and information in order for their requirements to be met. In order for a decision to undertake some activity to be morally responsible it is not sufficient to point to evidence suggesting that this activity brings some type of benefit and that no evidence exists in favour of serious risks being imposed by this activity. There also has to be positive evidence of sufficient quality to the effect that the activity in question does in fact not produce risks serious enough to make it a morally irresponsible course of action.

As regards the aspect of value neutrality, while it is certainly the case that the ideal of precaution presupposes some definite set of basic values to operate on, it can still concur with the basic tenet of the enlightment that there are no given experts on what these values are. Adopting the precautionary ideal is consistent with acknowledging this secular and anti-totalitarian aspect of the enlightment. Thus, in practice, a precautionary policy coheres with the idea of democracy as the final practical solution to how disagreements on what values should ultimately be socially pursued should be resolved. As mentioned, I will return to this issue below.

In all, then, rather than constituting a break with modernity and the enlightment ideals, the ideal of precaution should be seen as an improvement that continuous this line of historical development. Contrary to being a call for a retreat to dark ages of totalitarian, pastoral and/or arbitrary policy making, the ideal of precaution demands even more of scientifically informed and well thought-through policies based on an increased openness as regards what basic values are to be pursued by society.

6.4.2 The Remaining Challenge of Values

In Chapter 2, it was pointed out that what set of basic values any version of PP or the requirement of precaution is seen to operate on may have great impact on its practical prescriptions. Later on, it was remarked that the ideal expressed by the requirement of precaution is not neutral with regard to these values, but rather presupposes a definite set of values that helps to clarify the content of key-notions such as risk, chance, harm and benefit. In addition, there are competing ideas as to how plausible values can be traded off against each other in defensible ways that also need to be determined. With the brief exception of some aspects of the discussion of the weight of evil,[52] these issues have been left open throughout the investigations undertaken in this book. And, I am sorry to say, due to their complexity they will have to remain open in this particular book. Nevertheless, some comments regarding how these issues can be tackled within a policy of precaution along the lines suggested above are called for.

There are two issues with regard to values that are of particular importance with regard to precaution. First, it needs to be decided what types of entities can be harmed, what makes for harms to these entities and, as a consequence of this, what possible events should be seen as risks at all. As mentioned in Chapter 2, one of the most challenging aspect of this issue is to consider the validity of various non-anthropocentric ideas that ascribe value to entities such as non-sentient organisms, species, ecosystems and landscapes. To what extent should various possible effects on such entities be seen as risks worth to be considered in their own right (i.e., not only as a function of possible effects on humans or other sentient beings) from a precautionary point of view? Second, if any such idea is found to be warranted, we need to decide how these values should be balanced against competing anthropocentric values. That is, if there is a conflict between the interests of humans and that of 'nature' (however this latter interest is defined[53]), how should these interests be traded off against each other in order to reach a defensible decision from a precautionary point of view?

Moreover, this second issue of trade-offs arises even if we consider only traditional anthropocentric values. This is most clearly seen in connection to the notion of an acceptable or decent mix of risks and chances introduced in Chapter 5. It is possible to build into this notion an idea to the effect that the risks and chances that various individual parties are exposed to cannot be traded off against each other in any way theoretically possible. For example, it is not obvious that very small benefits (or chances) for each of a very large number of people can add up to a sum of benefits capable of balancing out severe harm to a few single individuals in a way that makes for an acceptable or decent mix of risks and chances. It is thus not built into my theory that the aggregation and balancing of individual risks, chances, benefits and harms has to conform to classic utilitarian formulas.

Now, if we look at actual policy making, issues such as these are in fact routinely addressed, namely when decisions are made that imply how different proposed interests and values are to be traded off against each other. So, I would suggest, political decision processes such as democratic ones are indeed capable of handling these types of problems. However, it is a severe shortcoming of most actual political processes that they are not very conducive to make the underlying issues of value salient. This problem of what may be called 'ethical transparency' is important, since, as was remarked in Chapter 2, how the value issues are resolved seems to

---

[52]The exception is one of the rigidity approaches – *rigidity of aggregation* – touched upon in Chapter 5.

[53]See Andersson (2007), for a recent exposition and critical discussion of this issue.

make a huge difference to what degree and price of precaution is prescribed by precautionary policies. Being a rather general problem of all political decision making, I have no 'cure' to suggest that would eradicate the lack of ethical transparency. However, I do believe that the model of precautionary policy making sketched above may serve to reduce this problem, since it presupposes that clear and publicly accessible guidelines for societal decisions on particular issues are developed.

6.4.3 The Case for Cosmopolitan Precaution

In the first section of this chapter, the point was made that the practical implementation of the morality of precaution put forward in this book first and foremost requires a politics of power. In connection to the issue of pollution and climate change, it was further remarked that the implication of my theory for the question of pollution trade seems to point to a basic structural problem with the standard bargain and agreement model of international policy making. Together, these remarks suggest that the morality of precaution provides a case for more robust, centralised and democratic policy making structures on the level of global politics – *cosmopolitanism*, as it is regularly called in political theoretical debate. To this may be added the general observations regarding the implications of my theory for issues of global distributive justice. In view of these, the negotiation and agreement model seems to give a disproportionate portion of power to affluent nations to continue their overproduction of (beneficial) risks, while less well-off nations are deprived of power to receive their fair share of defensible risk production. In contrast to this, a democratically designed centralised system of cosmopolitan policy making at least opens the door for nations with large populations but weak economic bargaining power to have a more substantial influence on the shape and content of global policy making in the areas of interest from the point of view of precaution.

In addition to this, there is a further argument that strengthens the tendency towards a support of cosmopolitanism from the point of view of precaution. This argument is best explained by revisiting the problems of levels and time-horizons that were briefly touched upon at the outset of Chapter 2, but then left aside through the main bulk of my investigation.

These problems are structurally similar, but are most easily explained regarding levels. The notion of levels refers to levels of policy making with regard to geographical scope. Should precautionary policies be applied at some regional level within nations, at national levels or at some international level, the upper extreme of which is the global level? The problem created by this issue is that, whatever precautionary policy we consider, if this policy is applied at a lower level than the global one, the aggregated result of the applications at lower levels may add up to a total mass of policies that fails to meet the requirements of the morality of precaution on a global level. This holds even if we assume the applications at the lower levels to be perfectly successful.

Now, by itself, the theory of the morality of precaution defended in this book does not imply this to be a problem. Indeed, one *could* argue that if robust policy making structures on the global level are lacking, there is no case for applying the ideas about the morality of precaution at this level (since there is no acting party for the theory to address).[54] However, I believe that the deeper moral convictions underlying the view on the morality of precaution defended in this book points in another direction. Rather than urging us to resign in the face of the lack of robust global policy making structures, they provide a reason for *creating* such structures. And the problem of levels provides substantial fuel for this reason.

The line of thought underlying this claim runs roughly as follows: The basic intuition underlying my theory of the morality of precaution is that creating risks always has a morally significant price. If there are no reasons in favour of the claim that paying this price is worthwhile, we should avoid creating risks as far as possible. Much of my discussion in earlier chapters has concerned what such reasons may consist in and what makes them stronger or weaker. However, implicitly, this discussion has also revealed factors that influence how morally serious the creation of a particular risk is. One of these factors is how this risk compares to the chances of benefits that are produced by the same activity producing the risk and how this mix of risks and chances compare to the similar mixes of alternative options. Four further factors have been suggested to influence the outcome of such comparisons: the moral seriousness of the outcome-, harm- or value-aspect of the risk, the likelihood-aspect of the risk, the quality of information or knowledge underlying the assessment of these two factors and, finally, the relative progressiveness idea of relating these factors to the idea of a decent mix of risks and chances. It is the first three of these factors that serve to add to the precautionary reasons in favour of cosmopolitanism.

First, when some type of harm occurs on a global scale it is generally more serious than when it occurs on lower levels. Second, as a result of this, the likelihood of bad outcomes to be very bad is greater on a global level than on lower levels. Third, at the same time, the basis of knowledge used for assessing these aspects of global risks is generally in a worse condition than the information underlying risk assessment on lower levels of policy making. All of this, I suggest, speaks strongly in favour of creating robust global policy making structures. First, these will be able to address those risks that are most important to address (and most irresponsible *not* to address properly) in both their outcome- and likelihood-aspects. Secondly, only such structures will be able to deal forcefully with the issue of improving the basis of knowledge used for assessing global risks.

On top of this, we may now add the problem of time-horizons. This problem is similar to that of levels, but with the difference that it concerns a totality of policies that fails to meet the requirements of the morality of precaution consisting of a temporal series (rather than a spatial collection) of policies that each taken by itself lives up to these requirements. In connection to this, it may suffice to remark that there are

[54]Cf. Rawls' argument that his theory of justice can only be applied to 'well-ordered' societies and therefore is inapplicable to global distributive issues (Rawls 1971).

strong reasons for suspecting the longest (and thus more serious from a precautionary point of view) such series to consist of collections of policies on a global scale (this since nations come and go, while a global collection of societies will preclude as long as there are enough humans capable of advanced social interaction).

In all, therefore, I conclude that a sound morality of precaution provides reasons for cosmopolitanism, which may be added to other such reasons.[55] Many of these other reasons connect to issues about global inequalities of distribution, unfair power balances in current systems of international politics, attaining the ability to deal with irrationalities of policy structures built on nation states as well as addressing in a forceful manner global threats such as large-scale environmental degradation (such as climate change). Moreover, all of these issues connect strongly to the classic cosmopolitan concern for securing world peace. So, I would expect the reasons I have been put forward here to fit well with the convictions of many cosmopolitan thinkers.

And here this book could have ended had it not been for the obvious fact that cosmopolitanism also has its possible downsides. I will conclude, therefore, on a more defensive note and argue that these can be dealt with in the context of policies aimed at meeting the requirements of the morality of precaution in ways preserving what is important in liberal democratic ideals.

6.4.4 Unrealistic and Dangerous?

This headline sums up the by far most common response to cosmopolitan ideas: it is indeed an appealing piece of utopia, but alas it will never be actualised. Working towards such a goal would thus be a waste of intellectual and material resources. Even worse, making such an attempt in spite of the lack of realism of the cosmopolitan vision would constitute a threat to international political stability and, therefore, world peace.[56] My own reaction to such criticism is threefold.

First, looking back into history and all the turbulences accompanying the development towards our current international political situation of a limited number of sovereign nation states, it is not clear to me that the cosmopolitan vision is unrealistic if only one adopts a sufficiently long time perspective. True, the path towards such a goal will be full of obstacles and, presumably, several concerned parties will have to pay a considerable price. However, at the same time, never before has there been such powerful reasons for individual parties (nations included) to accept substantial costs in terms of immediate self-interest in order to escape a situation that will be clearly undesirable from the perspective of us all. Since such global threats

[55] There is an immense literature on this subject, which has been debated at least since the seventeenth century. In this context, I refer the reader to a few works which are useful both through their intellectual content and/or their provision of references for further study: Archibugi et al. (1998), Beitz (1979), Dower (2003), Held (1995), Heater (1996), Hinsley (1963), Pogge (2008), Rawls (1999), Singer (2002), and Tännsjö (2008).

[56] An influential expression of this sort of criticism can be found in Rawls (1999).

to the prerequisites for everyone's existence and flourishing are far from declining, this pressure to find an escape-route can be expected to increase.

Second, in light of the just said, it seems to me that the most compelling reason to doubt that the cosmopolitan vision will ever see the light of day is if many people do actually doubt this. That is, the opinion that cosmopolitanism is an unrealistic idea runs an obvious risk of working mainly as a self-fulfilling prophecy. At the same time, this very insight has the potential of functioning as a tool for transforming such doubt into hope – thus undermining the accusation against cosmopolitanism of being unrealistic.

True, it still holds that *full* realisation of the cosmopolitan vision is a very ambitious goal, the road to which is so bordered by potential disturbances that it may appear *over* ambitious. However, what has been argued for above is merely that cosmopolitanism seems to be a good idea with respect to certain particular policy areas. These areas, in turn, have the kind of feature that we find in other areas where nations have actually surrendered some of their sovereignty in order to protect widely shared interests (rules of war and trade, and conventions of human rights, for example). Better still, the areas of most concern from a precautionary point of view actualise even more basic prerequisites for whatever other interests any nation may have, and may therefore very well be seen as particularly fitting for a first step in the cosmopolitan direction.

6.4.5 A Challenge for Liberal Democracy?

The ideal of liberal democracy aims at combining three partly conflicting political values: the value of an efficient politics of power to the benefit of all, the value of popular control over this politics of power and the value of the liberty of the individual. In the liberal democratic ideal, each of these values restrains each other through pulling partly in different directions. The value of efficient politics may, if unrestrained, support a gruesome totalitarianism but is unable to do so if the other two values are to be recognised. The value of popular control, if unrestrained, may support very inefficient systems of unanimous decision making on more or less every issue, but is tempered by the other values to settle for some system of majority decisions that imposes the will of the majority on individuals within certain limits. The value of individual liberty may, if unrestrained, reject almost any idea of a political power deciding what individuals are to do, but is tempered by the other two values to settle for a restriction on the extent to which a political power is allowed to influence the activities and choices of individual people. The outcome of balancing these tensions is the familiar system of (mostly) representative democracy where political decisions should not limit the freedom of people to do as they please as long as they are not exposing others to serious enough harm or peril.

The suggestions made above on how to apply my ideas about the morality of precaution to political decision making may be taken to be a threat to each of these three components of liberal democracy. However, the first of these possible threats has already been dealt with – namely the perception that precaution requires policies that would paralyse societal progress and thereby threat the value of efficiency. Already in Chapter 2, such versions of the ideal of precaution where dismissed as implausible. The morality of precaution suggested in this book is quite consistent with – indeed it demands – plenty of room for policy making to the benefit and progress of all.

What about the second value of popular control, then? Are not the ideas put forward above about cosmopolitanism, increased centralisation and long-term policy making at least in potential conflict with this part of liberal democracy? There are two reasons for believing so. First, the increased centralisation of cosmopolitanism would certainly move political power farther away from ordinary people in the sense that the weight of each individual voice in public debate and vote in elections would count for less relative to the entire political arena. Second, the request for longer time perspectives in policy making would seem to require that popular influence (such as shifts of government due to democratic elections) should not be allowed to upset or otherwise interfere too much with precautionary policy structures.[57]

As regards the first point, it should first be remarked that it may just as well be applied against any level of political decision making. As soon as more people join to form a democratic society, each individual will have to sacrifice some influence compared to if his or her society had remained a more regional or local organisation. But, at the same time, the resulting political unity which may be popularly controlled grows in size and this means that each individual will actually gain in influence through being able to have his or her say regarding issues that would otherwise remain entirely outside the sphere of her democratic control. When considering cosmopolitanism and the type of issues of most relevance from the point of view of precaution, this is especially salient regarding the vast number of people that today have absolutely no influence whatsoever over political decisions (or combinations of such) that greatly affect their lives, for instance, with regard to climate change and other sorts of pollution. In all, therefore, in the realm of precautionary policy making, cosmopolitanism would considerably increase the extent to which global policy may be popularly controlled in a sense of obvious relevance from a democratic point of view.[58]

The second argument may appear to be even more easily dismissed, since it may be claimed to be based on a misunderstanding of the message of the precautionary ideal. What is implied by the morality of precaution is that political decisions should have a certain content (i.e. prescribing a proper degree and price of precaution). This does not imply that it should not be possible to make decisions that go against this recommendation. In particular, it in no way suggests that the tools of democracy may not be used for searching out the best way of implementing the ideal

---

[57] Recently, this has led Ingmar Persson (2008) to argue that liberal democracy itself constitutes a major hurdle for effective global as well as national environmental policy.

[58] The basic connection between democracy as a political ideal and the extent to which people have an opportunity to influence political decisions affecting themselves is explored and defended in Arrhenius (2005).

of precaution in practical policy making. What it suggests is merely that if democratically decided policies do not observe the ideal of precaution, these policies suffer from a serious flaw. That is, even if democracy should be upheld, democratic decisions may be ill-considered, irrational or immoral. This response, however, needs to be developed in light of Persson's picture of a fundamental opposition between effective global environmental policy and popular control over societal governance.

What has been implicitly argued for above is that the morality of precaution pinpoints important features of policy that reside, so to speak, between the issue of what general procedures of policy making should be applied in society and the issue of what ultimate values and/or particular projects should be pursued by society. When settling for democracy with regard to the former issue, this does not mean that the question of whether society should be democratic or not is constantly on the political agenda. On the contrary, the common model is to make whatever solution to this issue that has been reached *extra resistant* to popular control by placing it in a certain section of the legal system commonly referred to as the constitution.[59] The latter issue, however, is allowed to be on the agenda in every phase of democratic policy making. Against this background, it may plausibly be suggested that certain procedural and formal aspects of precautionary policy making may be fitting for inclusion into the constitutional part of the legal system (while other parts having more to do with what ultimate values should be promoted and protected by such policy making remain on the regular democratic agenda). This suggestion gains in plausibility if the constitution allows for different degrees of resistance to popular control. In that case, the idea may be to view the constitutional parts of precautionary policy making not as equivalent to the most protected parts of the constitution, but at least as less easy to revise on the basis of popular vote than the day-to-day political issues. From a democratic point of view, this is acceptable, as long as it is possible to change the status of these aspects of precautionary policy by democratic means.

This takes us to the third value of individual liberty. Here, the threat against liberal democracy that may be perceived is somewhat different in that it connects more intimately to core features of the very idea of a morality of precaution. As sketched above, the traditional liberal paradigm is that people should be left free to do as they please, as long as they do not seriously enough harm others. In its traditional form, as presented by John Stuart Mill, this included the right of society to interfere with the freedom of individuals not only in cases where they actively harm someone, but also when they refrain from participating in social activities that are important for the well-being of all members of society (such as paying one's taxes).[60] But the typical view of Mill and other liberals is that these exceptions still

[59]There are, however, considerable differences between democratic countries as to the nature of such a constitution (in some cases it is not even written down) and the possibilities of changing or amending it through democratic means. For details, see, e.g., the *International Constitutional Law (ICL)* web resource.

[60]Mill (1982).

leave plenty of room for people to pursue their own happiness as they see fit without any need for societal involvement except for the protection of their right to do so.

However, if this view of things is supplemented with the idea of the morality of precaution suggested in this book, this room for individuals to follow their personal preferences may seem to shrink drastically. For not only will society be allowed to restrict individual action to prevent harm or secure the common good, it will also be so allowed in order to secure that a proper degree of precaution is upheld within society as a whole.[61] This, as we have seen earlier, involves not only preventing people from exposing each other to grave risks but also other things, such as restricting various practices due to want of sufficiently informative knowledge about risks and benefits and the securing of a morally acceptable level and distribution of risk-production in light of the relative progressiveness idea. Avoiding this sort of consequences seems impossible if the morality of precaution is to have any practical significance. The key question, therefore, is whether this expanded room for justifiable societal coercion and invasion of privacy can be justified on grounds acceptable from a liberal democratic point of view. My final point in this book is that it can.

As just described, already in its traditional form, the liberal ideal accepts societal coercion within certain limits. These limits are set according to a combination of certain moral values (that harm should be avoided, other things being equal) and a distinction between oneself and others ascribed political significance. Accordingly, while harming oneself is perhaps as morally serious as harming someone else, the use of societal force can only be justified when harm to others is at stake. The main point of my argument is that it is this latter aspect of liberalism that is essential for liberal democracy and that this aspect is in fact unmoved by the amendment to liberal theory of a morality of precaution. What this amendment affects is rather the first aspect: what is to count as such harm that, if it is imposed on others by someone, may justify the societal restriction of this person's liberty.

Of course, more traditionally inclined liberals may want to argue that there is something wrong with expanding the conception of harm to others in this way. However, if they are to do so they will, I conjecture, have to question the fundamental idea that imposing risks on others is morally significant in its own right. If so, they may as well question the idea that *harming* others is morally significant in itself. For, although different in various ways, one basic lesson of the investigations undertaken in this book is that the moral sentiments underlying traditional ethical ideals and the ethics of risks are as firm and worth taking seriously as one another. It is this insight that makes room for a morality of precaution capable of prescribing political action that should be as acceptable to liberal democrats as more familiar societal endeavours.

---

[61] This perceived threat to liberal values is stressed by Sunstein (2005, chapter 9).

References

Agar N. *Liberal Eugenics: In Defence of Human Enhancement.* London: Blackwell, 2004.
Ahteensuu M. "In Dubio Pro Natura? A Philosophical Analysis of the Precautionary Principle in Environmental and Health Risk Governance." PhD diss., Turku: Department of Philosophy, University of Turku, 2008.
Andersson P. "Humanity and Nature: Towards a Consistent Holistic Environmental Ethics." PhD diss., Göteborg: Acta Universitatis Gothoburgensis, 2007.
Archibugi D., D. Held, and M. Köhler, eds. *Re-imagining Political Community: Studies in Cosmopolitan Democracy.* London: Polity Press, 1998.
Arrhenius G. "The Boundary Problem in Democratic Theory." In *Democracy Unbound: Basic Explorations I*, edited by F. Tersman. Stockholm: Department of Philosophy, Stockholm University, 2005.
Ashcroft R.E. "Fair Process and the Redundancy of Bioethics: A Polemic." *Public Health Ethics* 1 (1) (2008): 3–9.
Axelrod R. *The Evolution of Cooperation.* New York: Basic Books, 1984.
Beck U. *Risk Society: Towards a New Modernity.* London: Sage.
Beitz C.R. *Political Theory and International Relations.* Princeton, NJ: Princeton University Press, 1979.
Boström N. and R. Roache. "Ethical Issues in Human Enhancement." In *New Waves in Applied Ethics*, edited by J. Ryberg and T.S. Petersen. Hampshire: Palgrave Macmillan, 2007.
Buchanan A., D.W. Brock, N. Daniels, and D. Wikler. *From Chance to Choice: Genetics and Justice.* Cambridge: Cambridge University Press, 2002.
Davis R.B. "The Principlism Debate: A Critical Overview." *Journal of Medicine and Philosophy* 20 (1) (1995): 85–105.
Dower N. *An Introduction to Global Citizenship.* Edinburgh: Edinburgh University Press, 2003.
Gibson D.G., J.I. Glass, C. Lartigue, V.N. Noskov, R.-Y. Chuang, M.A. Algire, G.A. Benders, M.G. Montague, L. Ma, M.M. Moodie, C. Merryman, S. Vashee , R. Krishnakumar, N. Assad-Garcia, C. Andrews-Pfannkoch, E.A. Denisova, L. Young, Z.-Q. Qi, T.H. Segall-Shapiro, C.H. Calvey, P.P. Parmar, C.A. Hutchison III., H.O. Smith, and J.C. Venter. "Creation of a Bacterial Cell Controlled by a Chemically Synthesized Genome." *Science* 329 (5987) (2010): 52–56.
Glover J. *What Sort of People Should there Be?* Harmondsworth: Penguin, 1984.
Glover J. "It Makes No Difference Whether or Not I Do It." In *Applied Ethics*, edited by P. Singer. Oxford: Oxford University Press, 1986.
Harris J. *Clones, Genes, and Immortality: Ethics and the Genetic Revolution.* Oxford: Oxford University Press, 1998.
Heater D. *World Citizenship and Government.* London: Macmillan Press, 1996.
Held D. *Democracy and the Global Order. From the Modern State to Cosmopolitan Governance.* Cambridge: Polity Press, 1995.
Hinsley F.H. *Power and the Pursuit of Peace. Theory and Practice in the History of Relations Between States.* Cambridge: Cambridge University Press, 1963.
Hobbes T. *Leviathan.* 1651.
Hunt G. and M.D. Mehta, eds. *Nanotechnology: Risk, Ethics and Law.* Sterling, VA: Earthscan, 2006.
Häyri M. *Rationality and the Genetic Challenge: Making People Better?* Cambridge: Cambridge University Press, 2010.
IPCC (Intergovernmental Panel on Climate Change). *IPCC Fourth Assessment Report: Climate Change 2007 (AR4),* Geneva: IPCC, 2007. Available online: http://www.ipcc.ch/publications_and_data/publications_and_data_reports.htm
Juth N. *Genetic Information: Values and Rights. The Morality of Presymptomatic Genetic Testing.* Göteborg: Acta Universitatis Gothoburgensis, 2005.

Malmqvist E. "Good Parents, Better Babies: An Argument about Reproductive Technologies, Enhancement and Ethics." PhD diss., Linköping: Linköping University, 2008.

Mayer J.E. "The Golden Rice Controversy: Useless Science or Unfounded Criticism?" *BioScience* 55 (9) (2005): 726–7.

Mayr E. *Systematics and the Origin of Species*. New York: Columbia University Press, 1942.

Mill J.S. *On Liberty*. London: Penguin, 1982.

Munthe C. "Preimplantation Genetic Diagnosis: Ethical Aspects." In *Encyclopedia of Life Sciences*. Chichester: John Wiley & Sons, 2007.

Munthe C. *Pure Selection: The Ethics of Preimplantation Genetic Diagnosis and Choosing Children Without Abortion*. Göteborg: Acta Universitatis Gothoburgensis, 1999c.

Munthe C. "Selected Champions: Making Winners in the Age of Genetic Technology." In *Values in Sport*, edited by C.M. Tamburrini and T. Tännsjö. London & New York: E&FN Spon, 2000.

Munthe C. *The Moral Roots of Prenatal Diagnosis: Ethical Aspects of the Early Introduction and Presentation of Prenatal Diagnosis in Sweden*. Göteborg: Centre for Research Ethics, 1996b.

O'Mathúna D. *Nanoethics: Big Ethical Issues with Small Technology*. London: Continuum International Publishing Group, 2009.

Palmer J. "Laser Fusion Test Results Raise Energy Hopes." *BBC News*, January 28, 2010. Available online: http://news.bbc.co.uk/2/hi/8485669.stm

Parfit D. *Reasons and Persons*, 2nd printing. Oxford: Clarendon Press, 1984.

Pennisi E. "Synthetic Genome Brings New Life to Bacterium." *Science* 328 (5981) (2010): 958–9.

Persson I. "Kan den liberala demokratin stoppa sin miljöförstörelse" [Can Liberal Democracy Stop Its Own Environmental Destruction?]. *Tidskrift för politisk filosofi* 12 (3) (2008): 7–26.

Peterson M. "The Precautionary Principle Is Incoherent." *Risk Analysis* 26 (3) (2006): 595–601.

Pogge T. *World Poverty and Human Rights: Cosmopolitan Responsibilities and Reforms*. Cambridge: Polity Press, 2008.

Rawls J. *A Theory of Justice*. Oxford: Oxford University Press, 1971.

Rawls J. *The Law of Peoples*. Cambridge, MA: Harvard University Press, 1999.

Resnik M.D. *Choices: An Introduction to Decision Theory*. Minneapolis & London: University of Minnesota Press, 1987.

Rummukainen M. "Reflections on the Uncertainty in Climate Scenarios." Paper presented at the conference Uncertainty and Active Risk Management in Agriculture and Forestry, Swedish Agricultural University, Alnarp, May 25–26, 2005.

Sandin P. *Better Safe than Sorry: Applying Philosophical Methods to the Debate on Risk and the Precautionary Principle*. Stockholm: Royal Institute of Technology, 2004.

Sandin P. and S.-O. Hansson. "The Default Value Approach to the Precautionary Principle." *Human and Ecological Risk Assessment* 8 (3) (2002): 463–71.

Shickle D. "Are 'Genetic Enhancements' Really Enhancements?" *Cambridge Quarterly of Health Care Ethics* 9 (3) (2000): 342–52.

Singer P. *One World. The Ethics of Globalization*. New Haven: Yale University Press, 2002.

Stork D. "22nd IAEA Fusion Energy Conference: summary of contributions on Fusion Technology and ITER Activities". Nuclear Fusion 49: 104002–104026, 2009.

Sunstein C.R. *Laws of Fear: Beyond the Precautionary Principle*. Cambridge: Cambridge University Press, 2005.

Tännsjö T. *Global Democracy: The Case for a World Government*. Edinburgh: Edinburgh University Press, 2008.

Tännsjö T. *Coercive Care: The Ethics of Choice in Health and Medicine*. London & New York: Routledge, 1999.

Tännsjö T. "Should We Change the Human Genome?" *Theoretical Medicine* 14 (1993): 231–47.

Tännsjö T. *Conservatism for Our Time*. London: Routledge, 1990.

United Nation's Framework Convention on Climate Change 1997, *Report of the Conference of the Parties on its third Session, held at Kyoto from 1 to 11 December 1997*. FCCC/CP/1997/7/Add.1, Bonn 1997: UNFCCC.

Index

C. Munthe, *The Price of Precaution and the Ethics of Risk*, The International Library of Ethics, Law and Technology 6, DOI 10.1007/978-94-007-1330-7,

MIX
Papier aus verantwortungsvollen Quellen
Paper from responsible sources
FSC® C105338
FSC
www.fsc.org

If you have any concerns about our products,
you can contact us on
ProductSafety@springernature.com

In case Publisher is established outside the EU,
the EU authorized representative is:
Springer Nature Customer Service Center GmbH
Europaplatz 3, 69115 Heidelberg, Germany

Printed by Libri Plureos GmbH
in Hamburg, Germany